NANOTECHNOLOGY APPLICATIONS

NANOTECHNOLOGY APPLICATIONS

K.P. Mathur

RAJAT PUBLICATIONS
New Delhi

Nanotechnology Applications
by *K.P. Mathur*

This edition is published by Rajat Publications
4740/23, Ansari Road Daryaganj,
New Delhi - 110002

ISBN: 978-81-7880-322-7

Edition 2026

Printed and bound in India

RAJAT
PUBLICATIONS

4740/23, Ansari Road, Daryaganj,
New Delhi- 110002 (India)
Mob:- +91-9811484973, +91-9811105389
E-mail: rajat_publication@yahoo.com

Preface

Nanotechnology today counts amongst the most interesting and exiting fields of science, which have emerged in the past decade . Embodying the adage small is beautiful nanotechnology is a form of micromechanics which is being used to develop working devices the size of a few nanometres, one month of a meter. This means that not only do our everyday appliances and devices are being rendered smaller, easier to handle, more efficient, it also means that 'nano' maybe also be rendered the order of the future world considering this, it is easy to discern the excitement that this field has generated.

Although a field still in its infancy, the relentless research and experimentation being carried out in it has enables US scientists to etch from silicon and electric motor that is smaller than 1 mm wide, and workable gears that have a diameter less than a human hair as further developments result in the manipulation of electrons and individuals atoms, the potential of nanotechnology is laid before us in its full glory—as a science which has the power to alter life as we know it Dramatically.

This book has been written so as to serve as an informative guide on the practical relevance of nanotechnology. What current nanotechnological services are in existence, on what principles they work, and what the future holds for nanotechnology—all these comprise the content of the books matter comprehensively concerned, all these issues facilitate a great understanding of nanotechnology. In addition, the meticulously detailed and label illustrations further foster a better understanding of nanotechnological devices. Lucid, easy-to-follow and compendious it is hoped that this book proves to be an illuminating reading and learning experience for the readers.

K.P. Mathur

Contents

1

Introduction

A revolution is occurring in science and technology, based on the recently developed ability to measure, manipulate and organize matter on the nanoscale — 1 to 100 billionths of a meter. At the nanoscale, physics, chemistry, biology, materials science, and engineering converge toward the same principles and tools. As a result, progress in nanoscience will have very far-reaching impact.

The nanoscale is not just another step toward miniaturization, but a qualitatively new scale. The new behavior is dominated by quantum mechanics, material confinement in small structures, large interfacial volume fraction, and other unique properties, phenomena and processes. Many current theories of matter at the microscale have critical lengths of nanometer dimensions. These theories will be inadequate to describe the new phenomena at the nanoscale.

As knowledge in nanoscience increases worldwide, there will likely be fundamental scientific advances. In turn, this will lead to dramatic changes in the ways materials, devices, and systems are understood and created. Innovative nanoscale properties and functions will be achieved through the control of matter at its building blocks: atom-byatom, molecule-by-molecule, and nanostructure-by-nanostructure. Nanotechnology will include the integration of these nanoscale structures into larger material components, systems, and architectures. However, within these larger scale systems the control and construction will remain at the nanoscale.

Today, nanotechnology is still in its infancy, because only rudimentary nanostructures can be created with some control. However, among the envisioned breakthroughs are orders-of-magnitude increases in computer efficiency, human organ restoration using engineered tissue, "designer" materials created from directed assembly of atoms and molecules, as well as emergence of entirely new phenomena in chemistry and physics. Nanotechnology has captured the imaginations of scientists, engineers and economists not only because of the explosion of discoveries at the nanoscale, but also because of the potential societal implications.

Nanotechnology's relevance is underlined by the importance of controlling matter at the nanoscale for healthcare, the environment, sustainability, and almost every industry. There is little doubt that the broader implications of this nanoscience and nanotechnology revolution for society at large will be profound.

Promises of Nanotechnology

Nanoscale science and engineering will lead to better understanding of nature; advances in fundamental research and education; and significant changes in industrial manufacturing, the economy, healthcare, and environmental management and sustainability. Examples of the promise of nanotechnology, with projected total worldwide market size of over $1 trillion annually in 10 to 15 years, include the following:

— *Manufacturing*: The nanometer scale is expected to become a highly efficient length scale for manufacturing once nanoscience provides the understanding and nanoengineering develops the tools. Materials with high performance, unique properties and functions will be produced that traditional chemistry could not create.

— *Electronics*: Nanotechnology is projected to yield annual production of about $300 billion for the semiconductor industry and about the same amount more for global integrated circuits sales within 10 to 15 years

— *Improved Healthcare*: Nanotechnology will help prolong life, improve its quality, and extend human physical capabilities.

— *Pharmaceuticals*: About half of all production will be dependent on nanotechnology affecting over $180 billion per year in 10 to 15 years.

— *Chemical Plants*: Nanostructured catalysts have applications in the petroleum and chemical processing industries, with an estimated annual impact of $100 billion in 10 to 15 years.

— *Transportation*: Nanomaterials and nanoelectronics will yield lighter, faster, and safer vehicles and more durable, reliable, and cost-effective roads, bridges, runways, pipelines, and rail systems.

— *Sustainability*: Nanotechnology will improve agricultural yields for an increased population, provide more economical water filtration and desalination, and enable renewable energy sources such as highly efficient solar energy conversion; it will reduce the need for scarce material resources and diminish pollution for a cleaner environment.

Scientific Understanding of Nature

The study of nanoscale systems promises to lead to fundamentally new advances in science and engineering and in our understanding of biological, environmental, and planetary systems. It also will redirect our scientific approach toward more generic and interdisciplinary research. Nanoscience is at the unexplored frontiers of science and engineering, and it offers one of the most exciting opportunities for innovation in technology.

Nanotechnology will provide the capacity to create affordable products with dramatically improved performance. This will come through a basic understanding of ways to control and manipulate matter at the nanometer scale and through the incorporation of nanostructures and nanoprocesses into technological innovations. It will be a center of intense international competition when it lives up to its promise as a generator of technology.

Nanotechnology promises to be a dominant force in our society in the coming decades. Commercial inroads in the hard disk, coating, photographic, and pharmaceutical industries have already shown how new scientific breakthroughs at this scale can change production paradigms and revolutionize multibillion-dollar businesses. However, formidable challenges remain in fundamental understanding of systems on this scale before the potential of nanotechnology can be realized.

Today, nanotechnology is still in its infancy, and only rudimentary nanostructures can be created with some control. The science of atoms and simple molecules, on one end, and the science of matter from microstructures to larger scales, on the other, are generally established. The remaining size-related challenge is at the nanoscale — roughly between 1 and 100 molecular diameters —

where the fundamental properties of materials are determined and can be engineered. A revolution has been occurring in science and technology, based on the recently developed ability to measure, manipulate and organize matter on this scale. Recently discovered organized structures of matter and new phenomena are scientific breakthroughs that merely hint at possible future developments.

The nanoscale is not just another step toward miniaturization, but a qualitatively new scale. The new behavior is dominated by quantum mechanics, material confinement insmall structures, large interfaces, and other unique properties, phenomena and processes.Many current theories of matter at the microscale have critical lengths of nanometer dimensions; these theories will be inadequate to describe the new phenomena at the nanoscale. Nanoscience will be an essential component in better understanding of nature in the next decades. Important issues include greater interdisciplinary research collaborations, specific education and training, and transition of ideas and people to industry.

Industrial Manufacturing

Nanotechnology is fundamentally changing the way materials and devices will be produced in the future. The ability to synthesize nanoscale building blocks with precisely controlled size and composition and then to assemble them into larger structures with unique properties and functions will revolutionize materials and manufacturing. Researchers will be able to develop material structures not previously observed in nature, beyond what classical chemistry can offer. Some of the benefits that nanostructuring can bring include lighter, stronger, and programmable materials; reductions in life-cycle costs through lower failure rates; innovative devices based on new principles and architectures; and use of molecular/cluster manufacturing, which takes advantage of assembly at the nanoscale level for a given purpose.

The Semiconductor Industry Association (SIA) has developed a roadmap for continued improvements in miniaturization, speed, and power reduction in information processing devices — sensors for signal acquisition, logic devices for processing, storage devices for memory, displays for visualization, and transmission devices for communication.

The SIA roadmap projects the future of nanoelectronics and computer technology to approximately 2010 and to 0.1 micron (100 nanometer) structures, just short of fully nanostructured devices. The roadmap ends short of true nanostructured devices because the principles, fabrication methods, and techniques for integrating devices into systems at the nanoscale are generally unknown. New approaches such as chemical and biomolecular computing, and quantum computing making use of nanoscale phenomena and nanostructures, are expected to emerge.

The molecular building blocks of life — proteins, nucleic acids, lipids, carbohydrates, and their non-biological mimics — are examples of materials that possess unique properties determined by their size, folding, and patterns at the nanoscale. Biosynthesis and bioprocessing offer fundamentally new ways to manufacture chemicals and pharmaceutical products.

Integration of biological building blocks into synthetic materials and devices will allow the combination of biological functions with other desirable materials properties. Imitation of biological systems provides a major area of research in several disciplines. For example, the active area of bio-mimetic chemistry is based on this approach.

Medicine

Living systems are governed by molecular behavior at the nanometer scale, where chemistry, physics, biology, and computer simulation all now converge. Recent insights into the uses of nanofabricated devices and systems suggest that today's laborious process of genome sequencing and detecting the genes' expression can be made dramatically more efficient through use of nanofabricated surfaces and devices. Expanding our ability to characterize an individual's genetic makeup will revolutionize diagnostics and therapeutics. Beyond facilitating optimal drug usage, nanotechnology can provide new formulations and routes for drug delivery, enormously broadening the drugs' therapeutic potential.

Increasing nanotechnological capabilities will also markedly benefit basic studies of cell biology and pathology. As a result of the development of new analytical tools capable of probing the world of the nanometer, it is becoming increasingly possible to characterize the chemical and mechanical properties of cells (including processes such as cell division and locomotion) and to measure properties of single molecules. These capabilities complement (and largely supplant) the ensemble average techniques presently used in the life sciences. Moreover, biocompatible, high-performance materials will result from the ability to control their nanostructure. Artificial inorganic and organic nanoscale materials can be introduced into cells to play roles in diagnostics (e.g., quantum dots in visualization), but also potentially as active components.

Finally, nanotechnology-enabled increases in computational power will permit the characterization of macromolecular networks in realistic environments. Such simulations will be essential for developing biocompatible implants and for studying the drug discovery process. An open issue is how the healthcare system would change with such large changes in medical technology.

Sustainability

Nanotechnology will lead to dramatic changes in the use of natural resources, energy, and water, as outlined in the following paragraphs. Waste and pollution will be minimized. Moreover, new technologies will allow recovery and reuse of materials, energy, and water.

Environment

Nanoscience and engineering could significantly affect molecular understanding of nanoscale processes that take place in the environment; the generation and remediation of environmental problems through control of emissions; the development of new "green" technologies that minimize the production of undesirable by-products; and the remediation of existing waste sites and streams. Nanotechnology also will afford the removal of the smallest contaminants from water supplies (less than 200 nanometers) and air (under 20 nanometers) and the continuous measurement and mitigation of pollution in large areas.

In order to hasten the integrated understanding of the environmental role of nanoscale phenomena, scientists and engineers studying the fundamental properties ofnanostructures will need to work together with those attempting to understand complex processes in the environment. Model nanostructures can be studied, but in all cases the research must be justified by its connection to naturally occurring systems or to environmentally beneficial uses. Environments for investigations are not limited and might include terrestrial locations such as acid mines, subsurface aquifers, or polar environments.

Energy

Nanotechnology has the potential to significantly impact energy efficiency, storage, and production. Several new technologies that utilize the power of nanostructuring, but developed without benefit of the new nanoscale analytical capabilities, illustrate this potential:

— Increasing the efficiency of converting solar energy into useful forn ..

— High efficiency fuel cells, including hydrogen storage in nanotubes.

— A long-term research program in the chemical industry on the use of crystalline materials as catalyst supports has yielded catalysts with well-defined pore sizes in the range of 1 nanometer to reduce energy consumption and waste; their use is now the basis of an industry that exceeds $30 billion a year.

— Developed by the oil industry, the ordered mesoporous material MCM-41, with pore sizes in the range of 10-100 nanometers, is now widely used for the removal of ultrafine contaminants.

— Several chemical manufacturing companies are developing a ¬anoparticle-reinforced polymeric material that can replace structural metallic components in automobiles; widespread use of those nanocomposites could lead to a reduction of 1.5 billion liters of gasoline consumption over the life of one year's production of vehicles, thereby reducing carbon dioxide emissions annually by more than 5 billion kilograms.

— Significant changes in lighting technologies are expected in the next ten years. Semiconductors used in the preparation of light emitting diodes (LEDs) for lighting can increasingly be sculpted on nanoscale dimensions. In the United States, roughly 20% of all electricity is consumed for lighting, including both incandescent and fluorescent lights. In 10 to 15 years, projections indicate that such nanotechnologybased lighting advances have the potential to reduce worldwide consumption of energy by more than 10%, reflecting a savings of $100 billion dollars per year and a corresponding reduction of 200 million tons of carbon emissions.

— The replacement of carbon black in tires by nanometer-scale particles of inorganic clays and polymers is a new technology that is leading to the production of environmentally friendly, wear-resistant tires.

Water

Global population is increasing while fresh water supplies are decreasing. The United Nations predicts that by the year 2025 that 48 countries will be short of fresh water accounting for 32% of the world's population. Water purification and desalinization are some of the focus areas of preventative defense and environmental security since they can meet future water demands globally. Consumptive water use has been increasing twice as fast as the population and the resulting shortages have been worsened by contamination.

Nanotechnology-based devices for water desalinization have been designed to desalt sea water using at least 10 times less energy than state-of–the art reverse osmosis and at least 100 times less energy than distillation. The critical experiments underpinning these estimations are underway now. This energy-efficient process is possible by fabricating of very high surface area electrodes that are electrically conductive using aligned carbon nanotubes, and by other innovations in the system design.

Agriculture

Nanotechnology will contribute directly to advancements in agriculture in a number of ways:

(1) molecularly engineered biodegradable chemicals for nourishing plants and protecting against insects;

(2) genetic improvement for animals and plants;

(3) delivery of genes and drugs to animals; and

(4) nano-array-based technologies for DNA testing, which, for example, will allow a scientist to know which genes are expressed in a plant when it is exposed to salt or drought stress.

The application of nanotechnology in agriculture has only begun to be appreciated.

Space Exploration

The stringent fuel constraints for lifting payloads into earth orbit and beyond, and the desire to send spacecraft away from the sun for extended missions compel continued reduction in size, weight, and power consumption of payloads. Nanostructured materials and devices promise solutions to these challenges.

Nanostructuring is also critical to the design and manufacture of lightweight, high-strength, thermally stable materials for aircraft, rockets, space stations, and planetary/solar exploratory platforms. The augmented utilization of miniaturized, highly automated systems will also lead to dramatic improvements in manufacturing technology. Moreover, the low-gravity, high-vacuum space environment may aid the development of nanostructures and nanoscale systems that cannot be created on Earth.

National Security

Defense applications include:

(1) continued information dominance through advanced nanoelectronics, identified as an important capability for the military;

(2) more sophisticated virtual reality systems based on nanostructured electronics that enable moreaffordable, effective training;

(3) increased use of enhanced automation and robotics to offset reduction^ in military manpower, reduce risks to troops, and improve vehicle performance;

(4) achievement of the higher performance (lighter weight, higher strength) needed in military platforms while providing diminished failure rates and lower life-cycle costs;

(5) needed improvements in chemical/biological/nuclear sensing and in casualty care;

(6) design improvements in systems used for nuclear non-proliferation monitoring and management; and

(7) combined nanomechanical and micromechanical devices for control of nuclear defense systems.

In many cases economic and military opportunities are considered to be complementary. Strong applications of nanotechnology in c'her areas would provide support for national security in the long term, and vice versa.

Marketing of Nanotech Products

Since economists have not yet really begun research on nanotechnology, their insights are somewhat tentative and based on experience with earlier technologies. A common paradigm is that new applications will be initially more costly than existing technologies, but will achieve better performance. However, completely new technologies may be cheaper, such as chemical manufacturing to mass produce nanoelectronic circuits as opposed to current methods using lithography in microelectronics.

Overall, nanotechnology will offer substantial advantages, being smaller, faster, stronger, safer, and more reliable. At the same time, it will require investments in new production facilities and in a host of ancillary industries supplying the raw materials, components, and manufacturing machines. Because it will take time to achieve economies of scale and to develop the most efficient fabrication methods, costs are likely to be relatively high in the beginning.

For this reason, nanotechnology-based goods and services will probably be introduced earlier in those markets where performance characteristics are especially important and price is a secondary consideration. Examples are medical applications and space exploration. The experience gained will reduce technical and production uncertainties and prepare these technologies for deployment into the market place. Similarly, in the private sector, technology transfer is likely to occur from performance-oriented areas to price-oriented ones. As a given technology matures, its cost may decline, leading to greater penetration of the market even where performance is not decisive.

The displacement of an old technology by a new one tends to be both slow and incomplete. Displacement of older methods will accelerate to the extent that nanotechnology extends its technical range and perhaps lowers its relative price. However, nanotechnology also is likely to stimulate innovations in older technologies that make them better able to compete — an ironic but potentially beneficial secondorder effect.

The diffusion and impact of nanotechnology will be partly a function of the development of complementary technologies and of a network of users. Whole new industries may have to be developed — along with the trained scientists and technicians to staff them.There may be many obstacles along this road that ordinary market processes cannot easily overcome. An important role for the government will be to invest in the long-term, high-risk, high-gain research needed to create these new industries and to ensure that they are consistent with broader societal objectives.

2

Drivers of Nanotechnology

Nanoscience and nanotechnology may turn out to have significant societal implications, as would be the case for any truly revolutionary advance in technology. We have identified three areas—natural resources, human condition, and security—where trends are raising significant social issues that will become drivers for technological change. To achieve a safe, secure world we must consider both global and national aspects of security, and the above issue areas are significant in this broader context. These problems are complex and require a life cycle systems approach for technological advances to contribute to real societal solutions. Finally, as with any radically new technology, the consequences of using nanotechnologies can harm as well as help mankind. It is up to society to debate and develop total and durable solutions.

The effect of nanotechnology on the health, wealth, and lives of people could be at least as significant as the combined influences of microelectronics, medical imaging, computer-aided engineering, and man-made polymers developed in this century. A government research and development initiative of this scale should go one step farther: the breakthroughs sought should relate to the central emerging problems of our society. While curiosity and unforeseen discoveries will still motivate the science, the scientific effort should point in the general direction of contributing elements to systems solutions to the complex challenges that face our nation. We need to think at an early stage about how nanotechnology will affect "the health, wealth, and lives of people."

National security cannot be independent of global security. But global security encompasses many more dimensions than just the military. The consequences of economic and informational globalization, combined with emerging demographic changes, will bring new kinds of threats to national and international security. Individual national security in a world of global collapse will not be tenable. We will have to seek national security in a context of global security. At our nuclear weapons laboratory, we tend to think of problems in terms of systems and life cycles. An example of a technology area where we as a nation did not work the entire life cycle problem is that of nuclear power. By not solving the nuclear waste disposal problem adequately as we developed the power generating systems, we left ourselves with a sizeable unresolved societal problem.

Our analysis of the conditions for global and national security leads us to consider three broad issues. The first revolves around the condition of the planet and its natural resources. We'll refer to this as the "green" issue involving foremost water, energy, and the environment. The second broad issue, which we call "red," is that of the human condition, with health at its center. These first two areas are potential sources of conflict that can drive global insecurity if unresolved. If it fulfills its promise, nanotechnology can enable solutions to many problems within these "red" and "green" issues areas. The third issue — which we call "black" — is military, for example as in the area of bio-warfare. Military advances enabled by nanotechnology, if used wisely in the interests of global security, can help to maintain a just peace. If used for purposes of aggression and domination, they can pose a substantial risk to all.

Natural Resources

There appears to be an increased potential for conflict as a rapidly growing world population tries to sustain itself with limited natural resources. The disparity in wealth between developed and developing nations, in combination with the uneven distribution of natural resources, remains a threat to the stability of states and of the international system.

With the advent of modern manufacturing, advanced technologies, and the information age, the importance of natural resources has been reduced for developed nations. But, especially for developing nations, the availability and control of critical resources such as oil, water, and food on an increasingly crowded planet remain among the major sources of long-term insecurity. It is possible that nanotechnology will contribute to easing resource disparities. Potential areas of impact include new materials, potable water, new energy sources, and sustainable environmental processes.

The availability of water resources remains one of the big issues for potential insecurity around the globe. As the World Commission on Water for the 21st Century has pointed out, What is obvious is that progress, especially in developing countries, is much too slow, and that unless there are drastic changes, water shortages and environmental degradation will become the norm. More people than ever will be added to some of the areas of the planet that are already most vulnerable socially, economically, and environmentally.

Low cost techniques for water purification, self-cleaning, evaporation reduction, and desalination could have tremendous impact by providing adequate supplies of clean water. A major driver for regional conflict might be removed. Adequate water supplies are necessary not only for human health, but also to assure the availability of food for the developing world's growing population. The potential impact of nanotechnology on water supplies is hard to predict at this time, but several areas of significant opportunity come to mind.

Affordable, engineered membranes that incorporated a self-cleaning process to avoid fouling could be used for large-scale desalination, which would go far in solving the water resource problem. While this technology would be a significant leap from current capabilities, the ability to tailor nanoscale membranes in combination with advances in self-assembly processes make it one to watch.

Another potential means of preserving water resources, particularly for agriculture, may be the control of evaporation through largescale application of nano-engineered films or membranes. Management of water resources is a good example of where the life cycle systems approach should be taken to assure that the technologies employed do not leave unanticipated environmental problems in their wake.

A second "green" issue of growing long-term concern for global security is that of energy resources and their use. Although proven reserves of oil and natural gas are large, heavy energy usage by the developed countries, combined with the demographic and development trends of the third world, will eventually put pressure on the supplies. In the meantime, the burning of fossil fuels has at least the possibility of substantially degrading the global environment.

Nanotechnology may be able to ameliorate energy problems both on the supply side and on the use side. In the near term, new, high-strength nanostructured magnets, nanolubricants, and other improved materials may greatly improve motor efficiency. Inthe long term, nanoengineered fuel cells, biocatalysts for crops for food or biomass fuels, or nanostructured photovoltaic films may permit cheaper alternative energy sources. For example, if the efficiency of photovoltaics were improved by a factor of two from the 20 to the 40% range at comparable costs — something that is theoretically possible — the role of solar energy would grow substantially. Likewise, if the oceans could be used for growing biomass fuels or harvesting energy through nano-biotechnology advances, significant increases in global energy supplies would result.

Systems life cycle thinking is particularly important in addressing the energy issue because of the coupling of energy and the environment. For example, if artificially engineered plants that produce ready-to-use energy become possible, at an early stage we will have to address issues akin to those now arising from the field of genetically engineered foods. But, properly designed, systems using such technologies as photovoltaics, engineered photosynthesis, factory process heat re-use, or agricultural fuel production could lead to a world of sustainable energy, agriculture, and climate. Such an "open system biosphere" would clearly have enormous implications for global security.

Nanoscience may also enable new materials and technologies that reduce economic dependence on other kinds of natural resources. The dependence of nations on extraction resources might be altered if common materials could achieve the functions of rarer and more costly materials. We refer to the ability to nanostructure a material for specific desired properties not found in its usual forms as *nano-alchemy*. In essence one is creating a new material by nanostructuring rather than by merely changing the chemistry. For example, common materials in the form of nanoscale clusters have been demonstrated to take on specific chemical catalytic properties, superior to those of more expensive catalysts. At this point we cannot predict whether nano-alchemy will apply broadly, or at what cost, but it is possible to envision large changes in how industrieswork. Then the relative wealth and power of nations could change, as could some of the contributing sources of international conflict.

There are no guarantees that the unregulated marketplace will assure a distribution of the benefits of nanotechnology that brings widespread prosperity and tranquillity. It is also possible to imagine the new technologies being used in ways that help the rich get richer and the poor get poorer. Given that nanoscience is being funded on a large scale from public resources, it behooves us to think on a national level about how its fruits can be directed to enhance national and international prosperity and security.

Human Condition

The human condition also must be considered in an analysis of the potential societal implications of nanotechnology. The proportion of the population at retirement age is increasing and will continue to grow rapidly over the next several decades, with a corresponding decrease in the available fraction of workers in the society. This trend has been strong in the developed countries where the birth rate has declined significantly, leading to low or even negative population growth, while life expectancy has been increasing. With an aging population, an increasingly large fraction of national and personal resources is being spent on health care. Here, we will not discuss the additional, very serious health issues, such as AIDS and other emerging infectious diseases that burden developing countries.

Desirable goals for an aging population include maintaining productivity longer, providing affordable health care, and deploying assistive technologies that maintain independence longer. Achieving these goals would greatly reduce the burden that an unhealthy, dependent older generation would place on younger citizens. The economic and social benefits to the nation would be great. We consider here just two possible connections to nanoscience and nanotechnology: assistive means to maintain physical independence and tools to support cognitive capability.

In the area of assistive devices, the ability to see (eye repair or hardware to replicate the eye function) and to maintain mobility (prostheses and sensor-based systems) could contribute significantly to maintaining productivity and physical independence. Advances in micro and nanotechnologies hold promise for contributing to a wide range of assistive solutions, from prosthetic limbs that adjust to the changes in the body, to more biocompatible implants, to artificial retinas or ears. Other opportunities lie in the area of neural prosthesis and the "spinal patch," a device envisioned to repair damage from spinal injuries.

In the area of cognition, revolutionary technical advances could have great impact on individual productivity and independence. We do not understand the workings of the brain well enough to predict with any confidence that assistive devices will actually work. However, rapid advances in the intersecting nano-, information science, andbiological sciences seem to promise significant surprises. Possible results include devices that enhance learning, cognition, judgement and decision making. Devices that helped people with dementia — nearly a third of the population over 85 — could have great impact. At the same time concerns about the use of artificial or assisted cognition for social control must be addressed.

As with the potential benefits of technologies relating to natural resource use, those relating to human health and quality of life also could end up being available only to small segments of the world's population. Today we talk of the "digital divide"; tomorrow it may be the "nano divide." Only the right combinations of public policy and free enterprise will lead to maximizing the societal benefits of the new technologies.

Security

There is little doubt that nanoscience and nanotechnology will carry implications for the use of force for military and civilian security. Military and police organizations would highly value enhanced situational awareness in a world of ambiguity, confusion, and asymmetric threats. The implications of advances in computing speed, higher density memories, enhanced sensing and communication, and microsystems that, individually or in swarms, may contribute to situational awareness and control are obvious. Nanoscience will enhance all of these technologies. Implications of such advances range from distributed early warning, assessment, and response systems, to enhanced decision support systems. New non-lethal weapons may also emerge.

One area in which our understanding is rapidly growing is that of the emergent behavior of collective systems. For example, researchers are beginning to appreciate how bees, with limited individual capabilities and simple rules of interaction, are collectively able to complete complex tasks, such as finding and harvesting nectar. Nanoscience, understanding of cognition, and microtechnologies may combine to give us small, smart devices that sense, think, act, and communicate as swarms. Robotic swarms might play important roles in both security situations and natural disasters where direct human presence would be dangerous or ineffective.

New information technology combined with better understanding of human and machine cognition, may give us new decision support systems. Information display and data fusion are already important military technologies. If memory aids can be integrated with the human brainfor decision support, applications in areas beyond military and emergency situations may become available. Related technologies would be interface devices such as wireless communication to the ear or displays on the retina, or reasoning support systems that would serve as decision advisors.

These various advances could contribute to global stability by enhancing the capabilities of peacekeepers to operate in difficult circumstances or of soldiers to resist aggression. As with other enhancements of military capability, however, they could also contribute to the success of military aggression. If the technologies were cheap and widely available, they could expand threats from terrorist or paramilitary groups.

Disruptive Technologies

"Disruptive technologies" are those which produce new products in new ways. Initially, they may cost more and be less effective than the more mature, "sustaining technologies." But eventually, they become so much cheaper and better as to drive the older technologies out of the market. The technologies emerging from nanoscience may well prove disruptive. If so, they will have societal implications that extend beyond their functional applications and into the realms of industry and economy. Particular manufacturing firms, and perhaps entire industries, might be deeply changed, or even shrink to insignificance. Some managers and workers might be put out of business, while others may prosper. Those with the resources and adaptability to retrain may succeed, while others — perhaps especially older workers — may not make the transition successfully. Redistributions of economic power could lead to corresponding redistributions of political influence.

The international status of the nations which first master the new technologies may rise, while the nations overly committed to old industrial processes or to extracted resources may fall behind. As on the national level, redistributions of global technological strength could result in realignments of global prosperity and influence. These changes could promote national and international stability and security — or they could hinder it.

Nanotechnology and Societal Transformation

Technological innovation sustains a fundamental tension of civilization — the tension between humanity's quest for more control over nature and the future, and our equally strong desire for stability and predictability in the present. Luddites were not against technology per se. They were against losing their jobs, and so they smashed the power looms that had put them out of work. The change wrought by technological advance continually remakes society, and this transformational process is on the one hand central to the dynamic that is commonly labeled "progress," and yet on the other is a source of continual destabilization and dislocation as experienced by individuals, communities, institutions, nations, and cultures.

In the age of science and technology (S&T), the government has increasingly become the prime catalyst for scientific advance and technological innovation. At the same time, modern government is also continually responding to and managing the transformational power of science and technology. Yet there is little effort to understand the relation between these two critical activities — advancing knowledge and innovation, and responding to their impacts — or to link them in a way that can enhance the value and capability of each.

A single technological innovation can remake the world. When the metal stirrup finally migrated from Asia to Western Europe in the 8th century, society was transformed to its very roots. For the first time, the energy of a galloping horse could be directly transmitted to the weapon held by the man in the saddle — a combat innovation of devastating impact. Because horses and tack were costly, they were possessed almost exclusively by landowners. Battlefield prowess and wealth were thus combined, and from this combination grew not just the traditions of a "warrior aristocracy" but the structure of European feudal society itself. Later, when the Anglo Saxon King Harold prepared to defend Britain against the invading Normans in 1066, he actually dispensed with his horse and ornamental wooden stirrups, choosing to lead his numerically superior forces on foot. The outnumbered Normans, however, boasted a strong, stirrup-equipped cavalry, and thus won the day — and the millennium.

Such narrative has the ring of mythology, yet the experience of the industrialized world reinforces the knowledge that a new machine can help change everything. The invention of the cotton gin in the late 18th century allowed a vast expansion of cotton cultivation in the American south — and directly fueled a commensurate rise in the importation and use of slaves for plantation labor. One hundred and fifty years later, the mechanical cotton picker suddenly rendered obsolete the jobs of millions of African American share croppers, and catalyzed a 30-year migration of five million people out of the rural south and into the cities of the north. While the development of the mechanical cotton picker was no doubt inevitable, its proliferation was consciously accelerated by plantation owners who, fearing the rise of the civil rights movement, sought quickly to find atechnological replacement for the existing system of exploitation labor upon which they were economically dependent.

These examples point not only to the power of new technologies to transform society, but to the comprehensive interconnectedness of technological change and the complex social structure of society. The invention of the stirrup as a battlefield tool was in some very intricate way connected to the development and expansion of feudalism in Europe; the evolution of agricultural technology for a single cash crop is indissolubly bound to the ongoing struggle to overcome the U.S. legacy of slavery, segregation, and bigotry. More familiarly, a single class of technology was a central determinant of geopolitical evolution after the end of World War II. Cars, television, air conditioning, and vaccinations have all stimulated foundational changes in society during the past century.

Of course new technologies rarely emerge in isolation. The industrial revolution is not just the story of harnessing steam power to factory production capability, but also the story of technological revolutions in transport, communication, construction, agriculture, resource extraction, and, of course, weapons development. These technological systems penetrated the innermost niches of society — home and family, school, workplace, community — and forced them to change. They also introduced completely new social phenomena, and stimulated the invention of completely new institutions.

The industrial revolution created the macroeconomic phenomenon of unemployment. Prior to the 19th century, even the most economically and politically advanced societies were dominantly agrarian and rural. For the majority of people, work was rooted in the home and the family. Vagaries of weather and transportation imposed irregularities and hardship, but most people and families harbored a diversity of skills that gave them independence from the marketplace and resilience to cope with a variety of challenges. In hard times, resort to subsistence farming and barter was usually possible.

Industrialization and urbanization linked workers far more closely to the larger economic market, while removing the need and ability for them to maintain the diverse skills necessary for survival in the pre-industrial world. The traditional connection between manufacturing and agriculture in the home was sundered by new economic organization and by geography. Labor itself became a commodity, subject to the same fluctuations and influences as other commodities.

During an economic downturn, factories fired people or closed down entirely. For the first time, workers could not easily respond to changing economic conditions by switching to a different type of work or moving to a subsistence mode. The political economist Karl Polanyi observed: "To separate labor from other activities of life and to subject it to the laws of the market was to annihilate all organic forms of existence and to replace them by a different type of organization, an atomistic and individualistic one".

As technological innovation interacts with society to create new phenomena, such as unemployment, society also responds by developing new types of institutions and response mechanisms. Today we can recognize the problem of unemployment as central to a diversity of social, political, and economic structures and activities ranging from theorganization of labor to insurance safety nets to educational programs to immigration policy. Unemployment rates are a key indicator of economic health, and a key determinant of political behavior.

National and international economic policies focus strongly on managing unemployment, even as theoretical investigations seek to clarify the relation between unemployment rates and other key attributes of modern economies. The general point is that transformational technology represents one variable in a complex assemblage of dynamic, interrelated societal activities. Decision making processes tend to address each of these activities in isolation from the others, e.g., conduct of research and development (R&D), dissemination of innovation products, devélopment of regulations, reform of institutions.

Concerted action occurs when a given innovation stimulates enough transformation to demand a response from other sectors of society. This response then triggers additional changes, which in turn demand further modulation. The process is reactive, discontinuous, disruptive, and sequential — like billiards. The challenge is to move toward a process of technology-supported societal progress where different sectors and activities can continually coevolve in response to knowledge about one another's needs and constraints — like an ecosystem.

A brief consideration of evolution of information technologies helps to bring this look at societal transformation into the present. Gutenberg's perfection of the printing press of course had enormous transformational impact, allowing the broad dissemination of written texts and consequent expansion of information

— and literacy — that undermined the Church's hegemony over knowledge and culture, and helped promote the dissolution of medieval social structure.

The implications of the information and communication revolution on democracy itself are far from clear. On the one hand, proliferation of information dissemination networks means greater access by more people to more information — and a greater capacity to communicate one's ideas and preferences in democratic fora. Control of information by authoritarian governments is becoming increasingly futile, and organization of democratic opposition increasingly enhanced, by new information technologies. But when this same capacity translates into 10,000 identical e-mail messages sent to a Member of Congress in support of a particular bill, one is hard-pressed to suggest that democracy is the beneficiary.

Of particular concern is the recent increase in public referenda aimed at bypassing the legislative process. The barriers to putting referenda on ballots have been enormously reduced by information and communication technologies that can be used to disseminate ideas and organize group action with relatively little effort. While on the one hand this type of direct democracy can be a refreshing antidoteto sclerotic legislative process, on the other it is quite often devoid of any serious deliberative process or public discourse, reflecting perhaps the pique of one wellorganized interest group or individual, and the substantiation of a Warholian politics where anyone with access to a decent list-serve can lead a movement for a day.

The implications of the information and communications revolution on the distribution of economic benefits in society are also problematic. Does the troubling increase in wealth concentration that characterizes the global economy derive from the way that advanced technologies diffuse in market economies? Does the synergistic character of information and communication networks mean that disenfranchised populations and nations will find it increasingly difficult to participate in the spectacular economic growth that we have seen in the past decade?

Society is illprepared to answer such questions, let alone act on them in a knowledgeable manner. Paradoxically, concerns about appropriability cut both ways. In the information society, the increasing ease of information dissemination may also threaten our system for protecting intellectual property and innovation.

We see the fingerprints of societally transforming technological systems in the controversy over genetically modified organisms; in the morally reprehensible situation where 24 million HIV-positive sub-Saharan Africans cannot possibly afford AIDS drugs that are widely available in the affluent world; in the existence of 40 million Americans with no medical insurance; in the general inability of our public school systems to create a citizenry able to take advantage of the opportunities of the knowledge economy; in the challenges presented by the aging of our population; in the rising atmospheric carbon dioxide levels that symbolize 150 years of industrial dynamism.

Even the unprecedented rise of civil and ethnic conflict throughout the world in the past decade can be plausibly connected to technological transformation. Approaching this phenomenon from entirely different directions, the political scientists Samuel Huntington and Benjamin Barber each conclude that advanced communication and informationtechnologies have created new fora for expressing ethnic identity and pursuing and strengthening cultural solidarity. Virtual communities, for example, can act to maintain identity over great distance, while also more efficiently garnering resources to support the expression of cultural goals.

Economic and Political Implications of Nanotechnology

The relationship between the meter and the nanometer can be understood better in monetary terms. A dollar bill is to a billion dollars, as a nanometer is to a meter. There are 1000 million dollars in a billion, and there are 1000 one thousand dollar bills in a million dollars. Thus a nanometer is 1000x1000x1000 times smaller than a meter, just as a dollar is 1000x1000x1000 smaller than a billion dollars. Nanotechnology deals with objects having at least one dimension in the range of nanometers, typically from 1 to 100 nm.

Scientists, from the onset of modern scientific thought in Greece, struggled for centuries to learn about the sizes of atoms and molecules. In 1905, Albert Einstein, as part of his doctoral dissertation, calculated the size of sugar molecules to be close to 1 nm. He used experimental results for the diffusion coefficient of sugar molecules in water measured by Graham, in the Einstein diffusion equation and obtained 0.99 nm as the answer. This historical development put the sizes of atoms and molecules on solid theoretical ground.

Recent developments in microscopy [such as transmission electron microscopy (TEM) and atomic force microscopy (AFM)] allow scientists today to see and manipulate particles of nanometer dimensions as easily as a child can manipulate with a needle grains of salt seen with a light microscope. The TEM and AFM microscopes are about one thousand times more powerful than the light microscope.

The transition from theory to practice, from calculating the size of very small objects to manipulating them at will, defines the onset of nanotechnology. The things that we can see and do when we deal with nanometer size objects are the subjects of nanotechnology. We can now leverage nano-scale particles into high-performance polymers and ceramics to yield composite materials having unique combinations of desired properties.

We can beneficially exploit these composites in useful applications that provide enhanced functionality and value. The automobile is one platform that is beginning to take advantage of nano-composites in diverse components and systems ranging from catalytic converters that more efficiently convert combustion byproducts to benign emissions, to economical light weight plastics and coatings that enhance fuel efficiency and vehicle durability.

Catalytic converters, found in most modern cars and in high efficiency gas burners and wood stoves, utilize nano-scale metal oxide ceramic coatings to efficiently present a highsurface area of precious metals like platinum and palladium to exhaust gases. These coatings are essential to speed and complete the conversion of harmful emissions such as carbon monoxide, nitrogen oxides, unburned hydrocarbons, and soot into benign byproducts such as nitrogen, water and carbon dioxide. This is a good example of the beneficial impact of nanotechnology on the environment and on the quality of life, especially in large cities.

Nano-particles derived from clays and related materials, strengthen and stiffen plastics. This property is being exploited by converting inexpensive, light plastics to engineered composites that have the unique combination of both stiffness and toughness. Such composites are being developed for use in automotive bumper systems, for exterior body panels, and for instrument panel structures and interior closeout panels. Nanoreinforcement enables the homogenization of plastic materials, which will facilitate recycling of plastics upon disposal of a vehicle.

The stiffening effect is also useful in the processing of plastics by increasing melt strength, which in turn enables the molding of larger and/or thinner parts without distortion. Blow molding — an inexpensive plastic manufacturing process — will especially benefit, enabling the production of lighter and less costly fuel tanks, bumper systems, and seating and other automotive and household plastic components. The use of polymeric nanocomposites in automotive applications will help to improve the efficiency of vehicles in miles per gallon and to reduce the volume of byproducts discharged to the environment.

It is hard to imagine nanotechnology without an intensive participation of the scientists and manufacturing technologies of the chemical industry. The assembly of atoms and molecules into materials and substances that are useful to society are essential to the high quality of life enjoyed by modern civilization.

Pharmaceuticals, plastics, electronics, textiles, food and many other things are the result of human ingenuity translated into useful and cost effective products by the work of persons skilled in the use of technology to convert raw materials found in nature into manufactured products. Nanotechnology can extend the quality and number of useful products made by industry because it can provide new building blocks and new tools to assemble them and to convert them into new products.

The size reduction of electronic circuits and components in the last forty years has resulted in the creation of new industries and in dramatic changes in our life styles. This incredible impact of size reduction in the electronics industry illustrates the potential of nanotechnology to change the future of all industries and the quality of life. The transition from the eye to the light microscope expanded our field of vision by about 1000-fold. Modern electronics circuits have components that are visible with the eye and the light microscope. We have another 1000-fold of magnification open to us from the light microscope to the TEM and the AFM. We can only imagine the potential that this 1000-fold factor will have in the evolution of electronic devices and in the creation of new technology — nanotechnology — in all industries.

Carbon nanotubes is a good example of the new generation of nano-materials to be produced by the chemical industry. They are far stronger than steel, but lighter, and conduct electricity like metals. New applications for these amazing nanotubes are being reported daily, in electronics, chemistry, optics, and biology. In the automotive and plastic industries they are being examined for their reinforcing capability, but also toimpart degrees of electrical conductivity to plastic composites. The utility of these features range from more efficient painting to potentially "plastic" circuitry integrated into insulating molded plastic articles.

Static dissipative plastics efficiently accept paint, reducing volatile emissions in the application of solvent born coatings and enabling the use of non-solvent-containing powder coatings on plastics. They are also desired in fuel system components to prevent static discharge fuel ignition. Semi-conductive plastics are desired as functional enclosures to shield electronic components from disruptive electromagnetic interference, which is becoming more of a challenge as the automotive industry migrates to new electrical architecture.

Universities, new start-up and established companies, are engaged in the inventions needed to produce carbon nanotubes and other nano-scale materials and products at economic costs. They are conquering the barriers involved in working with very small objects. The rules and theories that govern these new forms of matter are the topic at the frontiers of science and technology. The challenges in dynamics, architecture, assembly, and fabrication of nanomaterials are reinvigorating all fronts of science: from engineering disciplines and physics, to synthetic organic and inorganic chemistries, from polymer and materials science and polymer fabrication, to electronics and biology.

Recent developments provide a glimpse of the new industries that will be created by the natural evolution of knowledge into technology. New mesoporous solids offer larger surface areas and larger pores available for catalysis, absorption and separations. These big brothers of the molecular sieves and zeolites hold promise for the design and development of new catalysts and devices able to hold larger molecules, and reaction intermediates, and to facilitate molecular events with the precision and elegance of natural enzymes. The marriage of micelles, owned by colloid chemists, with silicate chemistry, has created a new window that is taking scientists to explore the interactions of other macromolecules such as proteins and enzymes with all types of substrates.

Combinatorial methods applied to bio-inorganic synthesis make it possible to find "wise" proteins and enzymes that grow crystals and materials replicating the elegance of coral growths, sea shells, or silicon chips. Thus, synthesis is moving from control of composition and structure to the control of size, shape, morphology and function. Soon we will have at our service the creative engines of nature to produce new materials with tools not very different to those used to make wine, cheese or beer.

Molecular electronic devices, such as redox switches, are proposed as components of new computer architectures to create chemically assembled electronic nano-computers from the "bottom up". Some of these devices are looking at DNA molecules as building blocks to fabricate circuits. This approach contrasts the "top down" technology used to manufacture electronic hardware today.

Quantum dots, semiconductor-based nanoparticles, are luminescent materials that can be used in all sorts of devices based on optical signal detection. Coupling these nano-electronic and nano-optical devices will create, a plethora of new technologies that will eventually displace the current "top down" built devices.

The chemical and automotive industries will be composers and musicians of the new harmonies to be produced by the evolution of nanotechnology. The creation of polymeric nanocomposites, new batteries, new electronic conductors, novel optical devices,catalysts and fuel cells, will lead to a transformation of the vehicle architectures into lighter, more efficient and high performance transportation products. The costperformance balance will be a key driver of this metamorphosis from conventional to nano-structured systems and components. Society at large will be the beneficiary.

Requirements for increased fuel economy in motor vehicles demand the use of new, lightweight materials — typically plastics — that can replace metal. The best of these plastics are expensive and have not been adopted widely by vehicle manufacturers. Nanocomposites, a new class of materials under study internationally, consist of traditional polymers reinforced by nanometer-scale particles dispersed throughout. These reinforced polymers may present an economical solution to metal replacement.

In theory, the nanocomposite can be easily extruded or molded to near-final shape, provide stiffness and strength approaching that of metals, and reduce weight. Corrosion resistance, noise dampening, parts consolidation, and recyclability all would be improved. However, producing nanocomposites requires the development of methods for dispersing the particles throughout the plastic, as well as means to efficiently manufacture parts from such composites.

Researchers at Mobil Oil Co. have revolutionized hydrocarbon catalysis by the development of innovative nanostructured crystalline materials. Their program focused on zeolites, porous materials with well-defined shapes, surface chemistry and pore sizes smaller than 1 nanometer. A new zeolite class, ZSM-5, was discovered in the late 1960s.ZSM-5 has a 10 atom ring structure that contributes pore sizes in the range 0.45 – 0.6 nm and enables shape selected chemistries not previously available.

Zeolite catalysts now are used to process over 7 billion barrels of petroleum and chemicals annually. New Zealand is using the same catalyst to produce 1/3 of its oil fuel requirement by converting it from natural gas via methanol and then high-octane fuels. ZSM-5, along with zeolite Y, now provide the basis for hydrocarbon cracking and reforming processes with a commercial value that exceeds $30 billion in 1999. Another example at Mobil Oil Co. is the aluminosilicate 10 nm shaped cylindrical pores, which have been applied in both catalysis and filtration of fine dispersants in the environment. Further systematic advances in nanotechnology are expected to increase its share of an overall world catalyst market that exceeded $210 billion in 1999.

Increasing ability to design and fabricate materials atom by atom has allowed creation of new materials with customized physical and electronic properties. An example of such a material is the amorphous alloy called Vitreloy™. The new material is twice as hard, twice as elastic, twice as strong, and twice as tough, compared to steel.

The rebounding and heat transfer properties are significantly different from the crystalline materials due to the different types of atoms and their arrangements. The atoms in the amorphous alloy Vitreloy™ are in a densely packed, but random arrangement. Amorphous materials are formed by cooling the liquid material quickly enough to prevent crystallization; the atoms do not have time to arrange themselves into an ordered structure. Because of the varying sizes of these atoms and their random arrangement in the solid, there are no groups of atoms that can easily move past one another. A consequence of this low atomic mobility is the low internal friction when a force is applied.

Information Technology based on Nanotechnology

The history of information technology has been one of learning to make "bits" smaller. There is no obvious and hard physical limit to the minimum size of logical devices that process information or the marks that store information. Indeed, quantum physics is being recast as a theory of information, and even a single atom can no longer be seen as the ultimate limit to the minimum size of a bit. Currently, the smallest logical devices being manufactured contain billions of atoms, and the smallest magnetic bits on commercial hard drives contain millions of atoms.

Assuming continued exponential improvement in our ability to pattern matter at ever-smaller dimensions, in perhaps 35 years we will have the capability to design and control the structure of an object on all length scales from the atomic to the macroscopic — in other words, the beginnings of a mature nanotechnology. To grasp some implications of a mature nanotechnology, imagine a world where information technology is truly ubiquitous and dirt cheap, where even trivial human artifacts contain extraordinary complexity and therefore extraordinary ability to process and communicate information. These broad capabilities of future information technology are easy to forecast, but their implications for society are still difficult to discern. History suggests that the most important future applications of the technology will surprise us.

Our present quandary over copyright law may seem quaint in a future where reproductions of any object are increasingly inexpensive and increasingly indistinguishable from the original. Privacy issues will be ever more important in a world where ever more of the objects around us share information with each other. Finally, while cost does not appear to be a factor that will limit broad access to the benefits of information technology, education will be the key to full participation in the economy, even more than it is today.

In order to have meaningful discourse on the societal impact of nanotechnology, we must first agree on what we mean by nanotechnology. In fact, definitions vary. This is to be expected, since there is no agreed-upon educational curriculum for someone who wishes to become a "nanotechnologist." No university offers an advanced degree in nanotechnology, although faculty are thinking hard about it. Nevertheless, a growing community of researchers are beginning to call themselves nanotechnologists. Members of this community come from such diverse disciplines as condensed matter physics, synthetic chemistry, materials science, biochemistry, and electrical and mechanical engineering.

Nanotechnology is about the creation and manipulation of information. Information is now understood to be a measurable, rigorously defined, fundamental construct of physics, on the same conceptual level as energy or entropy. Roughly speaking, the information content of a physical system is defined as the number of bits in its most concise description.

A perfect crystal has very little information content, since its structure can be described very concisely. All that is needed is a short string of bits to list the coordinates of silicon atoms to form a unit cell, and some bits to indicate repetition of the cell indefinitely to fill space. A perfect crystal is perfectly monotonous and therefore perfectly useless. But, if we supply the information needed to carve a particular pattern of impurities, metals and insulators into that crystal, it can become, for example, a microprocessor.

A typical microprocessor is one of the most complex, therefore information-rich, artifacts yet designed and built by our species. If we examine one closely, we find a nested hierarchy of structures. Some structures are as large as the entire silicon crystal (chip) and some structures consist of layers of differentiated materials only a few atoms thick. Integrating structure and function on many length scales down to the atomic, silicon microelectronics is one of the few existing examples of a true nanotechnology.

But if we measure all the structures in a microprocessor chip, we find the vast majority of dimensions to be on the order of 180 nanometers or greater. 180 nanometers happens to be the smallest principal dimension that can be routinely defined by the lithographic processes currently used in manufacturing. The smaller dimensions are achieved through controlled deposition of very thin films of material and a limited set of processing "tricks" which take advantage of forces at the atomic scale to help assemble the atoms. The amount of information we can incorporate into the structure is very limited at dimensions below 180 nanometers. Still, a microprocessor contains a lot more information than does a perfect crystal of silicon.

Now consider any living thing. Again we find a complex and nested hierarchy of structures, but now the hierarchy extends all the way to the sub-nanometer scale. Exchanging just two atoms in a molecule of DNA or RNA can make all the difference. A very long string of bits is required to describe even a single living cell. Indeed, the density of information in living things approaches the maximum that is physically possible. Furthermore, living things do not just contain information, but they continuously process vast amounts of information.

3

Societal Implications of Nanotechnology

The marriage of science and technology beginning in the latter part of the 19th century accelerated the process of innovation, and thus the process of societal transformation as well. If the industrial revolution played itself out in less than 200 years, the electronics revolution seems likely to have a working life of perhaps 75 years, while the biotechnology revolution, although hardly yet on its feet, is already prophesied to be supplanted by the nanotechnology revolution in the first half of the new century. What type of transformations might this revolution have in store? Our point here is not to predict the future of nanotechnology and its impacts — an impossible goal — but to illustrate the direction and scale of thinking that will be necessary if we are to successfully manage the interaction of new knowledge and innovation with society.

The promise of nanotechnology to remake our world seems virtually infinite. So the first thing to say is that if — as is variously claimed — nanotechnology is going to revolutionize manufacturing, health care, travel, energy supply, food supply, and warfare, then it is going, as well, to transform labor and the workplace, the medical system, the transportation and power infrastructure, the agricultural enterprise, and the military. Each one of these technology-dependent sectors is operated by and for human beings, who act within institutions and cultures, according to particular regulations, norms, and heuristics, all of which may reflect decades or even centuries of evolution and tradition. Not one of them will be "revolutionized" without significant difficulty.

The current chaos in our medical system is emblematic of this type of difficulty. In the near term, the current state of knowledge may suggest that the first wave of useful nanotechnologies will lie in the area of detection and sensing. The capacity to detect, precisely identify, and perhaps isolate single molecules, viruses, or other complex, nanoscale structures has broad application in such areas as medical diagnosis, forensics, national defense, and environmental monitoring and control.

When detection outpaces response capability — as it usually does — ethical and policy dilemmas inevitably arise. For example, it is already possible to identify genetic predisposition to certain diseases for which there are no known cures, or to diagnose congenital defects in fetuses for which the only cure is abortion. In the environmental realm, new technologies that detect pollutants at extremely low concentrations raise complex questions about risk thresholds and appropriate remediation standards. Thepresence of tiny amounts of toxic materials in groundwater may justifiably raise alarm among the public even if the health risk cannot be assessed, and the technological capacity for remediation does not exist. These types of dilemmas may be expected to accelerate and proliferate with the advance of nanodetection technologies.

Advances in sensing and detection may transform existing societal mechanisms and institutions that were designed to cope with uncertainty and incomplete or imprecise information. The insurance industry, for example, deals with incomplete knowledge about the health of specific individuals by spreading its risk among large populations. If there is no way to distinguish between someone who is going to suffer a potentially lethal middle-age heart attack, and someone who is going to live to 105, then they can both get health and life insurance. Society clearly gains from this arrangement: costs are broadly disseminated, and benefits are delivered to those who most need them.

Medical sensors that can, for example, "detect an array of medically relevant signals at high sensitivity and selectivity" promise to aid diagnosis and treatment of disease, but also to develop predictive health profiles of individuals. Today, health and life insurance companies often use pre-existing conditions as a basis for denying or restricting coverage. The advent of nanodetection capabilities will considerably expand the information that insurance companies will want to use in making decisions about coverage. The generation of new information might thus destabilize the risk-spreading approach that allows equitable delivery of social benefits to broad populations.

Nanotechnology offers a dizzying range of potential benefits for military application. Recent history suggests that some of the earliest applications of nanotechnology will come in the military realm, where specific needs are well articulated, and a customer already exists. One area of desired nano-innovation lies in the "increased use of enhanced automation and robotics to offset reductions in military manpower, reduce risks to troops, and improve vehicle performance."

Such considerations are simple extrapolations of current trends in technological innovation and societal transformation. More adventurous speculation is tempting but is perhaps best confined to science fiction novels. The question of public response to nanoinnovation, however, should not be avoided, even at this early stage. The ongoing experience of public opposition to old technologies such as nuclear power, new technologies such as genetically modified foods, and prospective technologies such as stem cell therapies, needs to be viewed as integral to the relationship between innovation and societal transformation.

Three observations are particularly relevant here. First, the impact of rapid technological innovation on people's lives is usually not consensual. Second, in the short term at least, the social changes induced by new technologies usually create both winners and losers. Third, rapidtechnological change can threaten the social structure, economic stability, and spiritual meaning that people strive in their lives to achieve. As the nanotechnology revolution begins to unfold in all its promise and diversity, such issues are bound to express themselves. They should not be viewed as threats, or as manifestations of intellectual weakness or repugnant ideology. Rather, they need to be recognized as a central part of the human context for technological change.

Preparing for Nanotech Revolution

Now nanotechnology had made nearly anything possible, and so the cultural role in deciding what should be done with it had become far more important than imagining what could be done with it. When resources are allocated for R&D programs, the implications for complex societal transformation are not considered. The fundamental assumption underlying the allocation process is that all societal outcomes will be positive, and that technological cause will lead directly to a desired societal effect. The literature promoting the National Nanotechnology Initiative expresses this view. The current policy approach thus addresses two elements:

— Conduct of Science and Technology

— Products of Science and Technology

These elements reflect the internal workings of the R&D enterprise. The fact that societal outcomes are not a serious part of the framework seems to derive from two beliefs:

(1) that the science and technology enterprise has to be granted autonomy to chose its own direction of advance and innovation; and

(2) that because we cannot predict the future of science or technological innovation, we cannot prepare for it in advance.

These are oft-articulated arguments, not straw men. Yet the first is contradicted by reality, and the second is irrelevant. The direction of science and technology is in fact dictated by an enormous number of constraints. And preparation for the future obviously does not require accurate prediction; rather, it requires a foundation of knowledge upon which to base action, a capacity to learn from experience, close attention to what is going on in the present, and healthy and resilient institutions that can effectively respond or adapt to change in a timely manner. If we flip the current S&T policy approach on its head, and start by thinking about desired social outcomes, rather than desired inputs to the R&D enterprise, where would we begin? We might identify several very general categories of outcomes that most people would agree are worth thinking about. For example:

— *Social equity:* the distribution of the benefits of science and technology.

— *Social purpose*: the actual goals of societal development that we want to pursue or advance.

— *Economic and Social enterprises:* the shape and make-up of the institutions at the interface between technology and the human experience.

The years since World War II have seen a very gradual evolution in the effort to connect thinking about S&T to thinking about the outcomes of S&T in society. A science policy report issued by the Truman Administration, for example, mentioned in its first pages the need to prepare for both the positive and negative impacts of scientific and technological change. The rise of the environmental movement in the late 1960s reflected a public demand that society devote more S&T resources to the achievement of desired social outcomes like clean air and water. The creation of the congressional Office of Technology Assessment reflected growing public concern about the need to understand the societal implications of technological choices.

Over the past decade, government-funded programs on the human dimensions of global climate change, and the ethical, legal, and social implications of the human genome project and information technologies, have been supported as adjuncts to much, much larger core research agendas in the "hard" sciences. Yet S&T policy itself remains input-driven.

Concepts such as sustainability, and analytical tools such as human development indicators, provide conceptual frameworks for linking R&D to societal outcomes, and in fact imply that outcomes are to some degree implicit in the choices we make about R&D inputs. These types of insights point the way toward the next step: to implement an approach to R&D policy that addresses the complex interconnections between technological advance and societal response. Such an approach would need to integrate the pursuit of innovation with an evolving understanding of how innovation and society interact, and include mechanisms to feed this understanding back into the innovation process itself.

If we wanted to be serious about preparing for the transformational power of a coming nanotechnology revolution, we would need first to get serious — at this very early stage—about developing knowledge and tools for more effectively connecting R&D inputs with desired societal outcomes. This in turn would require the creation of a dedicated intellectual, analytical, and institutional capability focused on understanding the dynamics of the science-society interface and feeding back into the evolving nanotechnology enterprise. Such a capability might include the following elements:

— *Analysis of past and current societal responses to transforming technologies.* A case history approach could be used to investigate the diverse avenues that society has followed in responding to a range of technological advances. Understanding the roles and relations between the media, academia, policy makers, institutions, and cultural factors could be the basis for assessing — and anticipating — the likely trajectories of technology-induced social change.

— *Comprehensive, real time assessment and monitoring of the nanoscience and nanotechnology enterprise.* At this relatively early stage, it should be feasible to build a database of important activities in nanotechnology, and then track the evolution of the enterprise over time, in terms of directions of research and innovation, resources used, public and private sector roles, publications and patents,marketed products, and other useful indicators. This type of information is essential to understanding potential impacts.

— *A science communication initiative, to foster dialogue among scientists, technologists, policy makers, the media, and the public.* Understanding, tracking, and enhancing the processes by which information about nanotechnology diffuses from the laboratory to the outside world is central to understanding the social transformation process as it occurs. Of equal importance is the need to understand and monitor how public attitudes and needs evolve, and how they reach back into the innovation system. Empirically grounded, research-based investigations on communication can be the basis for strategies to improve social choice in ways likely to secure favorable outcomes.

— *A constructive technology assessment process, with participants drawn from representatives of the R&D effort, the policy world, and the public.* Technology assessment is both a process for bringing together a range of relevant actors, and an evolving product that can inform and link the innovation and decision-making processes. Understanding the changing capabilities of both the nanotechnology enterprise and various sectors and institutions likely to be affected by the enterprise can contribute to a healthy policy making environment where innovation paths and social goals are compatible and mutually reinforcing.

Should nanoscience and nanotechnology yield even a small proportion of their anticipated advances, the impacts on society will be far-reaching and profound — "as socially transforming as the development of running water, electricity, antibiotics, and microelectronics". We can allow these transformations to surprise and overwhelm us, and perhaps even threaten the prospects for further progress. Or we can choose to be smart about preparing for, understanding, responding to, and even managing the coming changes, in order to enhance the benefits, and reduce the disruption and dislocation, that must accompany any revolution.

Nanotechnology and Societal Interactions

New technologies come into being through a complex interplay of technical and social factors. The process of innovation that will produce nanotechnology and diffuse its benefits into society is complex and only partially understood. Economists, as well as scholars in other fields, have long studied the generation, diffusion, and impact of scientific and technological innovation. These studies outline the variables likely to determine the rate and direction of these impacts, and to identify relevant research questions. They provide a foundation on which to build studies of societal implications of nanotechnology.

Scientific discoveries do not generally change society directly; they can set the stage for the change that comes about through the confluence of old and new technologies in a context of evolving economic and social needs. The thorough diffusion of even major new developments rarely happens all at once. Nanotechnologies are so diverse that their manifold effects will likely take decades to work their way through the socio-economic system. While market factors will determine ultimately the rate at which advances in nanotechnology get commercialized, sustained support for nanoscience research is necessary in this early stage of development so as not to become a rate-limiting factor.

Second-order Consequences

Perhaps the greatest difficulty in predicting the societal impacts of new technologies has to do with the fact that once the technical and commercial feasibility of an innovation is demonstrated, subsequent developments may be as much in the hands of users as in those of the innovators. The diffusion and impact of technological innovations often depends on the development of complementary technologies and of the user network. As a result, new technologies can affect society in ways that were not intended by those who initiated them. Often these unintended consequences are beneficial, such as spin-offs with valuable applications in fields remote from the original innovation.

For instance, consider how the Internet has progressed from a technology supported by the Department of Defense's Advanced Research Projects Agency (DARPA) to facilitate digital communications among universities with DARPA contracts, to a means by which teenagers and college students exchange music files. In another example, intended benefits may also have unintended or "second-order consequences."

Nanotechnologybased medical treatments, for example, may significantly improve life span and quality of life for elderly people; a second-order consequence would be an increase in the proportion of the population that is elderly, which might require changes in pensions or health insurance, an increase in the retirement age, or a substantial increase in the secondary careers undertaken by older people.

Another potential consequence that would need to be addressed is the potential increase of inequality in the distribution of wealth that we may call the "nano divide." Those who participate in the "nano revolution" stand to become very wealthy. Those who do not may find increasingly difficult to afford the technological wonders that it engenders. One near-term example will be in medical care: nanotech-based treatments may be initially expensive, hence accessible only to the very rich. Other consequences are not so desirable, such as the risk of closing old industries and environmental pollution, which sometimes becomes a problem, especially for largescale technologies.

To assess a nanotechnology in terms of its unintended consequences, researchers must examine the entire system of which the technology is a part through its entire life cycle. As the case of electric automobiles illustrates, without a careful analysis of the entire set of activities that produce, operate, and eventually dispose of a technology, people may leap to false conclusions about the extent to which the technology pollutes. For example, manufacture and disposal of an electric vehicle's battery may release more lead into the environment than if the vehicle had been fueled throughout its working life by leaded gasoline.

One concern about nanotechnology's unintended consequences raises the question of the uncontrolled development of self-replicating nanoscale machines. A number of very serious technical challenges would have to be overcome before it would be possible to create nanoscale machines that could reproduce themselves in the natural environment. Some of these challenges appear to be insurmountable with respect to chemistry and physical principles, and it may be technically impossible to create self-reproducing mechanical nanoscale robots of the sort that some visionaries have imagined. A new form of life different from that known would be a dramatic change that is not foreseen in the near future.

Initially, the impact of nanotechnology will likely be limited to a few specific products and services. Nanotechnology-based goods and services will probably be introduced earlier to those markets where consumers are willing to pay a premium for new or improved performance. Such primary effects would be to make things work better, cheaper, with more features, etc. This might, for example, increase food yields, generate new textiles for clothing, improve power production, or cure a certain disease.

By and large, the displacement of an old technology by a new one tendsto be both slow and incomplete. As a result, nanotechnology will coexist for a long time with older technologies rather than suddenly displacing them. During that time it will affect the further development of those competing technologies. Other secondary effects might be shifts in demand for products and services, so that people come to expect different kinds of food, medical care, entertainment, etc. This shift in demand may also initiate a tertiary effect, the need for augmented nanotechnology infrastructure — interdisciplinary research centers, new educational programs to supply nanoscientists and nanotechnologists, etc. Other tertiary effects would move upstream in our social structures and cultural patterns, such as shifts in education and career patterns, family life, government structure, and so forth.

An effective and cost-efficient way to protect the public and deal with nanotechnology's potential negative consequences is to develop a tradition of social-science-based countermeasures — and to support research in publicly recognized institutions on the processes that develop nanotechnology and apply it in diverse areas of life.

Ethical Issues

An important aim of a societal impact investigation of nanotechnology is to identify harms, conflicts over justice and fairness, and issues concerning respect for persons. For example, changes in workforce needs and human resources are likely to bring benefits to some and harm to others. Other examples of potential issues include safeguards for workers engaged with hazardous production processes, equity disputes raised by intellectual property protection, and questions about relationships between government, industry, and universities.

Scientists and engineers bring to their work a laudable concern for the social value of their labors. However, those working in a particular technical field may be focused on the immediate technical challenges and not see all of the potential social and ethical implications. It is important to include a wide range of interests, values, and perspectives in the overall decision process that charts the future development of nanotechnology. Involvement of members of the public or their representatives has the added benefit of respecting their interests and enlisting their support.

As scientists or dedicated scholars in their own right, they can respect the professional integrity of nanoscientists and nanotechnologists, while contributing a fresh perspective. Given appropriate support, they could inform themselves deeply enough about a particular nanotechnology to have a well-grounded evaluation. At the same time, they are professionally trained representatives of the public interest and capable of functioning as communicators between nanotechnologists and the public or government officials. Their input may help maximize the societal benefits of the technology while reducing the possibility of debilitating public controversies.

In addition, attention needs to be given to the individual responsibility of engineers, scientists, and others involved in the processes of generating powerful new nanotechnologies. Professional societies have a role to play in providing opportunities for discussing and devising guidelines that incorporate relevant ethical principles into emerging issues. Perhaps most importantly, ethics must be incorporated effectively into the curriculum for training new nanoscientists, nanotechnologists, and nanofabrication technicians.

Education of Nanotechnologists

The modern world faces the daunting challenge of attracting enough of the best graduate students to the physical sciences and engineering disciplines. Under present conditions, far too few good students are attracted to the fields relevant to nanotechnology. To some extent, this is a problem faced by all of the sciences, but the problem is especially acute for nanotechnology because a very large number of talented scientists, engineers, and technicians will be needed to build the nanotechnology industries of the future, and these professionals will require an interdisciplinary perspective.

Development of nanotechnology will depend upon multidisciplinary teams of highly trained people with backgrounds in biology, medicine, applied and computational mathematics, physics, chemistry, and in electrical, chemical, and mechanical engineering. Team leaders and innovators will probably need expertise in multiple subsets of these disciplines, and all members of the team will need a general appreciation of the other members' fields. Developing a broadly trained and educated workforce presents a severe challenge to our four-year degree and two-year degree educational institutions, which favor compartmentalized learning. Because current educational trends favor specialization, there must be fundamental changes in our educational systems.

However, introducing new degree programs in nanotechnology that provide a shallow overview of many disciplines, none in sufficient depth to make major contributions, may not give students the training that is needed to meet the future challenges. The right balance between specialization and interdisciplinary training needs to be worked out through innovative demonstration programs and research on the education process and workforce needs.

Education in nanoscience and nanotechnology requires special laboratory facilities that can be quite expensive. Given the cost of creating and sustaining such facilities, their incorporation into nanotechnology workforce development presents a considerable challenge. Under the present education system, many engineering schools, let alone the two-year-degree colleges, cannot offer students any exposure to the practice of nanofabrication.

Innovative solutions will have to be found, such as new partnerships with industry and the establishment of nanofabrication facilities that are shared by consortia of colleges, universities, and engineering schools. Web-based, remote access to those facilities may provide a powerful new approach not available previously. Despite the tremendous educational challenges, the exciting intellectual, economic, and social opportunities of nanotechnology might become a major factor in reinvigorating our nation's youth for careers in science and technology.

A related educational challenge is the very small number of social scientists who have the technical background and research orientation that would allow them to conduct competent research on the societal implications of nanotechnology. At the university level, liberal arts education gives far too low a priority to scientific literacy.

Social science professional societies, universities, and government agencies will have to make a long-term commitment to attract talented young social scientists to this area of research and to encourage them to gain the necessary professional skills and awareness of nanotechnology. This will require research on the societal implications of nanotechnology at a consistent and high enough level to establish this as a viable field of social science research.

Social Science Research Approaches

It is important to have social scientists study the processes by which nanoscience is conducted and nanotechnology is developed — even at this early stage. The knowledge gained will help policymakers and the public understand how nanoscience and nanotechnology are advancing, how those advances are being diffused, and how to make necessary course corrections. Insight into the innovation process will also grow.

Social scientists and scholars possess many effective ways of studying the development of new technology and its implications for society. Some methodologies suitable for studying nanotechnology are known; others will have to be identified or developed. Ethnographic techniques, such as those traditionally employed by anthropologists, are appropriate for some of this work. Also useful will be interviews of research and development teams, conducted over time and augmented by surveys and historical methods, to document the evolution of the knowledge and technology. Interviews, social network techniques, studies of communication patterns, and citation analysis of publications more generally can offer insights into the diffusion of scientific discoveries and ideas. Application of a scientific idea to a technical problem, technology transfer, and introduction of products into the marketplace can be tracked through statistics on research and development investments, patent applications, and new products and services.

With concerted effort, it will be possible to develop a number of indicators that provide early signs of change. One challenge to social science research will be to identify "bellwethers," "early adopters," or "first movers." For example, some geographic areas and strata of society experience technological change earlier than others do. Incipient transformations may reveal themselves first in start-up companies, university labs, and Internet communications.Innovating activity takes place in academe, industry, government laboratories, Federal agencies, and professional societies. For each group, measures and methodologies for studying the process and content of change must be developed.

In academe, key indicators might include interdisciplinary work, new courses, fellowships, information flow, and regional coalitions. In the private sector, key indicators might include investments, startups, and corporate partnerships. For government laboratories, data would cover budgets, equipment, standards, and coalitions, and for agencies, examination might be given to new initiatives, databases, and centers. The professions will create some number of new forums, symposiums, journals, and job fairs where interdisciplinary topics and careers would flourish. Social science research areas relevant to the process of discovery, invention, and development include appearance of new ideas and innovations, change in societal goals, and shift in commercial investment.

The societal impacts of nanotechnology may be of great scope and variety. A second research challenge is to address both short- and long-term impacts, intended and unintended, and first-order through Nth-order consequences. Because nanotechnology presents a highly diverse set of novel technical possibilities, accurate prediction of even the immediate consequences of individual innovations may be impossible. Some impacts will be surprising, and others will have emergent implications that will reveal themselves only over a long period of time. Ultimately, both technology and society are elaborate systems with the potential for chaotic and variable feedback mechanisms. Nanotechnology has such promise to impact so many aspects of society that predictions will be uncertain and difficult to validate empirically. This observation should not discourage researchers, however, but should inspire them to invest considerable sophistication and effort in their work. The domains and measures of potential social impacts include: economic growth, employment statistics, social transformations and medical statistics.

A third challenging, but important, area for social science research is the social acceptance, resistance, or rejection of nanotechnology. Representative sample surveys, supplemented by focus groups and open-ended interviews, can measure affective, cognitive, and psychosocial parameters. In recent years, political scientists and sociologists have developed new computation-intensive techniques for studying coverage by the news media; they have been tested in research on public controversies and are ready to track the changing public perceptions of nanotechnology. These and more traditional methods can also be applied to charting the process of regulatory review and approval, court decisions that actively sanction the use of the technology, mobilization of political support and opposition, and the activities of relevant social movements.

There are multiple feedback loops in which society responds to new innovations and in so doing transforms the context in which innovation occurs. As more and more new nanotechnologies are publicized and actually appear in the marketplace, the variable degree of social acceptance will become ever more important. Indicators to measure social acceptance of nanotechnology will be needed in the following areas: economic, political, religious and cultural.

Institutional Infrastructure

Nanotechnology's vast scope and the necessity of bringing together researchers from different disciplines may require that some of the important social science research be carried out by large teams, housed in research centers set up for this purpose. At the same time, many of the research methodologies require social scientists to be on site where nanoscientists, nanotechnologists, and decision-makers are doing their work. One model that meets both of these requirements is the virtual distributed research center (VDRC). Under this approach, each VDRC would be organized around a specific but somewhat broad set of scientific questions and research methods, so that the members would have a common framework for designing, carrying out, and communicating their research.

To ensure that results reflect the wide diversity of nanotechnology, social scientists would have to examine a range of empirical settings — for example, by conducting ethnographic research in a variety of nanotechnology laboratories. Thus many individual members of the VDRC would be situated, or would spend large blocks of time, at the geographically dispersed sites where they are studying. However, the VDRC would have a physical center that coordinates the work, develops and maintains funding and institutional partnerships, and supports effective communications among the far-flung team, both electronically and face-to-face in periodic meetings.

Many important nanotechnology-related questions could best be examined by more traditional centers, teams, and individual investigators. For example, survey research on public attitudes might best be done by a conventional team of researchers connected to one of the existing social survey organizations. Some research on economic trends, changing labor markets, and publication patterns could be done by individual investigators with access to data already available. Finally, there will always be a need for innovative projects carried out by individual scientists or small teams to develop new theories and methodologies and to carry out reconnaissance studies of emerging social phenomena.

Recommendations for Nanotech Development

Given the tremendous potential benefits of nanotechnology, and the concern that it be developed with sensitivity to potential negative implications, the workshop participants offered the following recommendations:

— Make support for social and economic research studies on nanotechnology a high priority. Include social science research on the societal implications in the nanotechnology research centers, and consider creation of a distributed research center for social and economic research. Build openness, disclosure, and public participation into the process of developing nanotechnology research and development program direction.

— Create the knowledge base and institutional infrastructure to evaluate nanotechnology's scientific, technological, and societal impacts and implications from short-term (3 to 5 year), medium-term (5 to 20 year), and long-term (over 20 year) perspectives. This must include interdisciplinary research that incorporates a systems approach (research-technology development-societal impacts), life cycle analysis, and real time monitoring and assessment.

— Educate and train a new generation of scientists and workers skilled in nanoscience and nanotechnology at all levels. Develop specific curricula and programs designed to:

(a) introduce nanoscale concepts into mathematics, science, engineering, and technological education;

(b) include societal implications and ethical sensitivity in the training of nanotechnologists;

(c) produce a sufficient number and variety of well-trained social and economic scientists prepared to work in the nanotechnology area;

(d) develop effective means for giving nanotechnology students an interdisciplinary perspective while strengthening the disciplinary expertise they will need to make maximum professional contributions; and

(e) establish fruitful partnerships between industry and educational institutions to provide nanotechnology students adequate experience with nanoscale fabrication, manipulation, and characterization techniques.

— Encourage professional societies to develop forums and continuing education activities to inform, educate, and involve professionals in nanoscience and nanotechnology.

— Involve social scientists at the onset of major nanotechnology R&D activities, while the technology is still in an early stage of development, from vision setting to development projects.

— Prepare corresponding management plans and policies to ensure that we can respond flexibly to implications as they appear on the horizon.

— Invest in significant new innovative efforts to educate and train the nanoscience and nanotechnology workforce, including the sensitivity to societal implications and the introduction of nanoscale concepts in mathematics, science, engineering, and technological education. Conduct a comprehensive study to determine the distinctive educational and workforce issues related to nanotechnology and seek potential solutions for problems that are identified in that study.

— Support interdisciplinary research that includes a systems approach (researchtechnology-society), life cycle analysis, and real-time monitoring and assessment.

— Study the evolution of disruptive technologies, the winners and losers in major technological transformations, and the implications for the economy. From those studies, project the social purpose, social equity implications, and social enterprise dynamics of anticipated nanotechnologies.

Recommendations to Organizations

Academe:

— Focus on multidisciplinary work on key research and education issues concerning socio-economic implications. Support interdisciplinary interactions between the physics, chemistry, biology, materials, and engineering communities on one hand, and the social sciences and economic science research communities on the other hand.

— Educate and train a new generation of scientists and workers for nanoscience and nanotechnology at all levels.

— Create local information centers for the public, teachers, industry, and scholars.

Private Sector:

— Provide intellectual input and seed funding of activities aimed at assessing the societal implications of nanotechnology.

— Develop partnerships with academic institutions and other sectors.

— Offer accessibility to social science researchers and provide feedback on societal implications studies.

Government R&D Laboratories:

— Establish interdisciplinary teams for major grand challenges in nanotechnology including socio-economic perspectives, including social scientists.

— Develop databases for evaluation and continuously update scenarios for the future.

— Establish user facilities available to industry and academe that enable integration of basic and applied research.

— Support nanotechnology research within laboratories emphasizing national defense mission.

Government Funding Agencies:

— Support nanotechnology researchers and social scientists to study the societal implications of nanotechnology.

— Support NNCO or an advisory group to monitor developments and examine the socio-legal implications of nanoscience and nanotechnology and take appropriate actions. Communicate the resulting activities to the public.

— Provide coordinated support for long-term basic research and shorter-term technological developments to create the technological base and prove the potential of the new technology.

— Establish dialog between NNCO and information technology (NSO) and biotechnology (BECON) coordinating groups elsewhere in the Federal Government.

Professional Societies:

— Develop forums and continuing education activities to inform, educate, and involve professionals and the public.

— Provide suggestions for grand challenges and suggest warning signs of potential risks.

Nanotechnology will fundamentally transform science, technology, and society. In 10 to 20 years, a significant proportion of industrial production, healthcare practice, and environmental management will be changed by the new technology. Economic growth, personal opportunities, sustainable development and environmental preservation will be affected. To take full advantage of the new technology, the entire scientific and technology community must involve all participants, including the general public; creatively envision the future; set broad goals; and work together to expedite societal benefits.

4
Environmental Applications

Environmental applications of nanotechnologies demonstrate their potential to revolutionise entire industries and displace major existing technologies. In spite of the few examples to the contrary, the costs and risks of transitioning from discovery to commercialisation are too great and the benefits too ill-defined for large companies to undertake the wide range of development projects required to bring nanotechnologies into practice. There is a need for government-sponsored mechanisms to facilitate this development.

An increased knowledge of the dynamics of processes specific to nanoscale structures in natural systems can improve understanding of complex processes occurring in the environment and lead to development of approaches for mitigating environmental harm. Collaboration between universities, national laboratories, and industry is also essential to the development of nanotechnology with application to the environment. This could be accomplished by establishing cross-disciplinary fellowships to support researchers for extended visits to institutions in the other sectors.

With the development and application of nanoscience comes the potential for great advances in processes and products, especially in the fields of medicine and manufacturing. However, both of these fields have direct effects on human health and the environment. Although nanosize particles exist naturally, this is the first time that they have been intentionally produced in large quantities to attain specific attributes. Once these nanoparticles are released into the environment, it is not known how easily then can be reclaimed or how they will bioaccumulate. The particles are known to cross the blood-brain barrier in humans and animals. In order to determine the safety of these particles, extensive testing and research will need to be undertaken under a number of different conditions and time frames.

In nanoscience, smart materials include all materials engineered at the nanoscale to perform a specific sophisticated task. Many of these materials are dynamic, meaning that they can change their basic structure or properties with the introduction ot an outside stimulus, such as self-tinting automotive glass that changes with amount of sunlight.

An environmental application of a smart material might be a self-cleaning membrane used for water purification. As more research takes place in this field, new technology develops that allows us to create materials based on molecular properties, macroscopic properties, and biological processes. However, with new technologies comes the need for new ways to assess potential hazards through all points in a product life cycle.

From an environmental standpoint, manufacture and ultimate disposal of these materials will bring the greatest risks and therefore will have the greatest need for close study and regulation. The use of nanotechnology as sensors is becoming one of its most notable advantages of nanoparticles because of their ability to respond in a recognisable way to the presence of a desired substance. This technology is geared towards the detection of many properties including temperature, water, light, sound, electricity, particular molecules, bacteria, toxins, explosives, and DNA. Sensors use the principles of molecular recognition in order to detect the presence of molecules. By developing sensors on the nanoscale, it allows for the detection of small traces of toxins and pollutants in the atmosphere, soil, and water.

The application of nanoscale science and technology to the more established field of biotechnology gives way to many new advances and research opportunities. The vast field of biomedical applications includes the creation of artificial bone, safe gene therapy, early cancer detection, and even the restoration of partial eyesight to the blind. Creating artificial bone relies on guided self-assembly of particles into the desired shape. The product is a bone substitute that is very strong and compatible with the human immune system and not disruptive to normal body functions. Applying nanoscience to gene therapy has the potential to make it a safer and more efficient process by compacting DNA into ultra-tiny balls, allowing DNA to be delivered to cells via liposomes. These would act as genetic faults and possibly curing diseases such as cystic fibrosis. Nano-sized particles hold great promise for drug delivery since they can be designed specifically to interact with particular molecules. These particles also aid in drug delivery throughout the body, as they are small enough to cross the blood-brain barrier. Delivering anti-tumor drugs this way, for example, could lead to new non-invasive ways to treat brain cancer.

Many of these applications and processes are very close to final stages of testing and approval, with countless more in earlier phases of research. Optobionics, a biomedical company based in the Chicago area, has patented a small chip packed with photodiodes for implantation into the retina in order to restore partial vision to those affected by retinitis pigmentosa. As with any new medical breakthrough, it is essential that these new products be thoroughly tested to determine long-term effects on health. Nanotechnology also has numerous applications in energy, with the aim of both making current techniques environmentally friendly and developing cleaner technologies such as solar panel and fuel cells.

New filtering systems for natural gas pipelines and smokestacks can be designed at the molecular level to remove even the most minuscule impurities, resulting in cleaner burning with fewer smog-creating impurities and more sensitive emission scrubbers. In addition, nanoparticles are increasingly being used to capture and convert solar energy more efficiently. These particles are being used in everything from panels to paints that can coat any surface and convert sunlight to electrical energy, including on the exterior of cars. Additionally, companies such as Sony, Mitsubishi, and Kodak are already beginning to substitute organic light emitting diodes (OLEDs) in lieu of widely used liquid crystal displays in many of their commercial applications due to the high efficiency of OLEDs, resulting in a longer lifetime.

Nanotechnology has many promising applications in remediation of existing environmental problems and proactive pollution prevention. Engineered nanospheres, resembling tiny molecular cages, can trap polychlorinated biphenyls (PCBs) and toxic metals, preventing them from flowing into the atmosphere and reducing their concentration once already introduced to the environment. Already zerovalent nanoiron is being used for in situ remediation of TCE at Superfund sites across the nation. Using an in situ remediation technique allows scientists to focus on the source of the contamination in a way that is not usually possible, and so far monitoring of the sites has not revealed any undesirable effects of the Fe nanoparticles in the soil.

New bottom-up fabrication processes will eliminate waste and allow for better breakdown of hazardous materials. Since we are building devices at the molecular level, products will be smaller, which means less waste will be produced resulting in less trash to dispose. On-site production, possible by molecular manufacturing, also eliminates the need for shipping, which is especially relevant for limiting the transport of potentially hazardous materials. Filtering systems for drinking or waste water can be designed at the molecular level to remove even the most minuscule impurities. In addition to improving air and water quality, nanoparticles can also act as catalysts for solar reactions in many applications and produce more efficient energy.

Nanometer-sized particles have been developed to improve the mechanical properties of tires, initiate photographic film development, and serve as vital catalysts in the petrochemical industry. To some extent, environmental scientists and engineers are already working with nanoscale structures. Natural weathering of minerals, such as iron oxides and silicates, and microorganisms, such as bacteria and algae, produce nanoscale colloids, which include dispersions of nanosized-particles in media with special properties that can be important in the fate, transport, transformation, and bioavailability of environmentally harmful substances. Anthropogenic and natural colloids of solids, liquids, and gases in liquids (sols, emulsions, and foams, respectively), and solids and liquids in gases (smokes and fogs, respectively) are commonly encountered in the environment.

Nanotechnology is about structure and the ability to work-observe, manipulate, and build-at the atomic or molecular level. This results in materials and systems that often exhibit novel and significantly changed physical, chemical, and biological properties due to their size and structure. The basic structures of nanotechnology include nanoparticles or nanocrystals, nanolayers, and nanotubes. These nanostructures differ in how they are made and how their atoms and molecules are ordered. A nanoparticle-a collection of tens to thousands of atoms measuring about 1-100 nm in aggregate diameter-is the most basic structure in nanotechnology. Such nanoparticles are created atom by atom, so the size, and often the shape, of a particle is controlled by experimental conditions. These particles can also be described as nanocrystals because the atoms within the particle are highly ordered, or crystalline. Nanostructures, are often arranged or self-assembled into highly ordered layers arising from hydrogen bonding, dipolar forces, hydrophilic or hydrophobic interactions, gravity, and other forces. Many naturally occurring biological structures, like membranes, vesicles, and DNA, form because of such self-assembly. Repeating structures with a tailored periodicity are essential in many applications of nanotechnology, such as photonics, catalysts, and membranes.

Understanding and building nanostructures through self-assembly is the core of the nanotechnology creation process. The extraordinary mechanical, electronic, and chemical characteristics of nanostructures called carbon nanotubes, which were discovered by Sumio Iijima in 1991, have captured the imagination of scientists and engineers. Nanotubes generally consist of hexagonal lattices of carbon atoms arranged spirally to form concentric cylinders. The tubes are nearly perfect crystals and thinner than graphite whiskers. Single-wall nanotubes (SWNTs) have a diameter of approximately 1.4 nm, and multiwall nanotubes (MWNTs) consist of between 2 and 30 concentric tubes that form an outer diameter of 30-50 nm. Nanotubes range in length from a few tens of nanometers to several micrometers. Nanotubes have characteristics that lead to various applications. For example, nanotubes are stronger than steel but very flexible and lightweight, which makes them a suitable material for both superstrong cable and tips for scanning probe microscopes. In addition to the remarkable mechanical properties, nanotubes could potentially replace copper as an electrical conductor or, with a small change in structure, replace silicon as a semiconductor.

Nanoscale Technologies in Pollution Control

Pollution prevention refers to "source reduction" and other practices that efficiently use raw materials, energy, water, or other resources to reduce or eliminate creation of waste. This strategy also includes using less toxic and renewable reagents and processing materials, where possible, and the production of more environmentally benign manufactured products. Nanotechnology could play a key role in pollution prevention technologies. For example, nanotechnology-based home lighting could reduce energy consumption by an estimated 10% in the United States, saving $100 billion annually and reducing carbon emissions by 200 million tons per year. Nanostructured catalysts, for example, can make chemical manufacturing more efficient by providing higher selectivity for desired reaction products. For example, aluminosilicate molecular sieves (zeolites) are porous crystalline solids with well-defined structures widely used for separations and catalysis. Nanometer-sized zeolites (10-100 nm) are being developed to selectively oxidize hydrocarbons, such as toluene to benzaldehyde. Using nanostructured zeolites makes this example more environmentally benign for two reasons. First, the oxidation reaction is initiated by visible light, which reduces energy consumption. Second, using visible light accesses low-energy reaction pathways that help eliminate wasteful secondary photoreactions and increase the yield of the desired product.

Assembling nanostructures from biopolymers or bio-inspired materials is an example of an environmentally benign approach to fabricating microelectronics. According to recent calculations, 1.7 kg of fossil fuel and chemicals and 32 kg of water are required to produce a single 2-gram, 32-megabyte microchip. Biomolecular nanolithography represents a "bottom-up" approach to replace current semiconductor chip production methods. Nanoscale metal particles on a biopolymer (poly-Llysine) template or scaffold are stretched out on a surface into well-defined chip architectures, such as lines and grids, at room temperature. Biodegradable materials like poly-L-lysine are used for construction of these structures.

Nanotechnology applications could also help create benign substances that replace currently used toxic materials. For example, nontoxic, energy-efficient computer monitors are replacing those made of cathode ray tubes (CRT), which contain many toxic materials. Newer liquid crystalline displays are smaller, do not contain lead, and consume less power than CRT computer monitors. Using carbon nanotubes in computer displays may further diminish the environmental impacts by eliminating toxic heavy metals and drastically reducing material and energy use requirements, while providing enhanced performance for consumer needs. Field emission displays (FEDs) that use carbon nanotubes are the latest development in display technology. Although the toxicity of carbon nanotubes is largely unknown, the amount of carbon nanotubes used in FEDs is quite small (0.5 g/monitor) compared to the kilograms of lead in CRTs.

Remediation and end-of-pipe treatment technologies

The early impact of nanotechnology research has been mostly in remediation and end-of-pipe treatment technologies. Various reports have appeared on the use of a variety of nanoparticles for treatment and remediation of pollutants in the environment. For example, nanoparticles of various oxidants, reductants, and nutrients have been suggested as useful for promoting contaminant transformation and stimulating microbial growth, because their small size and larger surface area make them more reactive and more flexible in terms of deployment. Nanoparticles can also exhibit unique chemical reactivity not observed for larger particles because of their unusual crystal shapes and lattice order. For example, conventional methods e.g., bioremediation and zero-valent iron) for in situ remediation of chlorinated organic solvents, such as trichlorethylene, tend to produce undesirable byprod-ucts, such as dichloroethylenes and vinyl chloride. Using nanoscale bimetallic particles essentially eliminates all the undesirable byproducts.

Nanoparticles that are activated by light, such as the large band-gap semiconductors titanium dioxide (TiO_2) and zinc oxide (ZnO), continue to be studied for their ability to remove organic contaminants from various media. These particles are readily available, inexpensive, and have low toxicity. Recently, ZnO nanoparticles were shown to act as both a sensor and photocatalyst for treatment of chlorinated phenols. Researchers are also quite interested in manipulating the surface of nanoparticles with organic or inorganic dyes to extend their photoresponse from UV to visible light, making them even more efficient as photocatalysts for the transformation of environmental contaminants because UV light represents only 5% of the solar spectrum.

Nanoparticles could provide enormous flexibility for in situ remediation as well. For example, nanoparticles are easily deployed in ex situ slurry reactors for the treatment of contaminated soils, sediments, and solid wastes. Alternatively, they can be anchored onto a solid matrix such as carbon, zeolite, or membrane for enhanced treatment of water, wastewater, or gaseous process streams. Direct subsurface injection of nanoscale iron particles, whether under gravity-feed or pressurised conditions, has already been shown to effectively degrade chlorinated organics, such as trichloroethylene, to environmentally benign compounds. The technology also holds great promise for immobilising heavy metals and radionuclides.

Research has demonstrated that nanoscale, bimetallic particles, such as iron/palladium, iron/silver, or zinc/palladium, can serve as potent reductants and catalysts for a large variety of common environmental contaminants such as PCBs, organochlorine pesticides, and halogenated organic solvents. Nanoscale metallic particles reduced virtually all chlorinated hydrocarbons tested to benign hydrocarbons. Additionally, ample evidence indicates that ironbased nanoparticles can be used to reduce many other recalcitrant contaminants, including anions (perchlorate, nitrate, and dichromate), heavy metals (nickel and mercury), and radionuclides (uranium dioxide). Nanotubes have been suggested as "a superior sorbent" for dioxins. The sorption energy of dioxin on carbon nanotubes is almost 3 times that of activated carbon. The Langmuir constant, a parameter characterising the sorption affinity, is many orders of magnitude higher for nanotubes than for activated carbon. Thus, the prospect of using carbon nanotubes for air and water pollution control appears to be very favorable, but large-scale applications of nanotubes in the near future are limited by cost and availability. Another example of environmental treatment and remediation-related application of nanomaterials includes using dendritic nanoscale chelating agents for polymer-supported ultrafiltration (PSUF). Dendrimers are highly branched polymers with controlled composition and an architecture that consists of nanoscale features. These nanostructures can be designed to encapsulate metal ions and zero-valent metals, enabling them to dissolve in suitable media or bind to appropriate surfaces.

Sensing and Detection

Rapid and precise sensors capable of detecting pollutants at the molecular level could greatly enhance our ability to protect human health and the environment. Manufacturing process control, compliance, ecosystem monitoring, and environmental decision making would be significantly improved if more sensitive and less costly techniques for contaminant detection were available. Of particular interest are remote, in situ, and continuous monitoring devices capable of yielding real-time information, and also those that can detect pollutants at very low concentration levels. Recently, chemical sensors based on individual SWNTs have been demonstrated. Altered electrical resistance of semiconducting SWNTs through exposure to gaseous molecules, such as nitrogen dioxide or ammonia, forms the basis for these sensors. The nanotube sensors also exhibited fast responses at room temperature to the gases and a substantially higher sensitivity than existing solidstate sensors.

Boron-doped silicon nanowires (SiNWs) have been used to create highly sensitive, real-time electrically based sensors for biological and chemical species. The SiNWs could be modified in various ways to provide different functionalities. Specifically, the amine- and oxide-functionalised nanowires were used to detect changes in pH, and biotinmodified SiNWs were used to detect streptavidin in the picomolar concentration range. In addition, antigen- functionalised SiNWs showed reversible antibody binding and concentrationdependent detection in real time.

Detection of the reversible binding of the metabolic indicator calcium was also demonstrated. These mechanisms could apply to de- tecting pathogens and chemical and biological agents in water, air, and food. Human health and ecological diagnosis could also benefit from sensors. For instance, nanobarcodes-highly encodable and chemically diverse nanoparticles -may allow for high-level multiplexing of small sample volumes. Evaluating the exposure of complex systems in the environment to toxic pollutants often has such requirements for sample analysis. Nanobarcodes can greatly enhance our ability to identify the source and strength of contaminants, determine the route and mechanism of environmental fate and bioavailability, and assess the effectiveness of treatment and remediation techniques.

Environmental and Energy Technologies

A number of environmental and energy technologies have already benefited substantially from nanotechnology.

Reduced waste and improved energy efficiency. Catalysis has also been demonstrated in some cases that catalytic efficiency (rate and turnover) is substantially altered by the use of nanoscale reagents in homogeneous or heterogeneous applications. Nanoscale materials for Li battery cathodes such as aerogel or zerogel V_2O_5 can substantially increase capacity, cell life, and chargedischarge rates.

Environmentally benign composite structures. The ability to incorporate nanoscale inclusions in composites has the potential to produce materials with improved properties and tailored to specific applications such as improved filtration systems. This can produce systems with increased environmental robustness, resulting in longer service life and reduced overall system costs and replacement needs, and reduced environmental impact. It also can produce lighter, smaller structures, resulting in systems with reduced energy consumption. Use of nanoscale materials in composites can range from the use of oxide- or nitride-based inclusions in steels to the development of fully engineered heterogeneous composites.

Waste remediation. Nanostructured materials have an increasingly important role in the remediation of wastes. This can take many forms, from using TiO_2 particles to oxidise organic contaminates, to employing nanoscale scavengers to capture heavy metals in contaminated waste sites. In many cases, illuminated particles can be oxidising agents, active either in solution or in aerosols. Recent work has shown that UV-illuminated nanoscale TiO_2 can be employed to clean atmospheric contaminates, including hazardous organic chemicals, cells, and viruses, as well as to sequester hazardous chemicals.

Nanoscale particles with suitable surface derivatisation can also be used to sequester heavy metals or bond to and passivate contaminated surfaces. In addition, it is clear that the more efficient chemical processes are, the less direct waste is produced. One interesting possibility is that as the surface chemistry of nanostructured materials is better understood, it will be possible to tailor the surface in nanostructured material-mediated reactions to minimise the generation of wastes. More recently, the discovery of the ordered mesoporous material MCM-41 has extended the pore size range that is attainable in structured inorganic materials to the 10 to 100 nm size ranges. It took nearly a decade to take these materials from the initial discovery to a commercial product. One of the uses of this product is in waste remediation of heavy metals at nuclear power sites. The ability to tailor the nanostructured materials for specific ions is important both in being able to remove all the waste and segregate it, and in lowering the cost of the process.

Energy conversion. Use of energy, indirectly as electricity and as fuel for transportation, is responsible for an enormous and adverse impact on the environment. Nanoscale systems offer the potential for renewable energy conversion systems with much less waste production; when this potential is coupled with improved batteries or fuel cells having nanoscale or mesoscale electrodes for energy storage in transportation, the positive impact on the environment could be tremendous.

Barriers and Solutions

Nanoscale processes

There is an urgent need to encourage interdisciplinary research that involves novel approaches and that adopts newly developed experimental, theoretical, and computational methods for characterising nanostructures. If interaction is increased between the community of scientists and engineers studying the fundamental properties of nanostructures and the community attempting to understand complex processes in the environment, this will hasten both communities gaining an integrated understanding of the environmental role of nanoscale phenomena. Overall, there exist opportunities in a number of areas:

— Replacement of waste-generating technologies with "green" technologies

— Improvement of process efficiency and manufacturing of smaller and lighter materials in order to reduce material and energy use

— Better understanding and control of natural phenomena and pollution through use of nanosensors and nanoelectronics

Nanoparticles and mitigation

The key to avoiding or remediating pollution by nanoparticles is understanding the basic science of the interaction of nanoparticles and nanostructured materials with the environment, especially with biological systems. Little is known of the key factors governing size, shape and surface chemistry. There are no effective predictive models. Before these are developed, the pollution potential of nanoparticles can only be assessed empirically, and in many cases that is quite difficult. Nanoparticles have the potential to be "micro-reactors" that can be employed in a wide variety of circumstances to convert energy, remediate waste, and provide sensing. There are a tremendous number of uses for these microreactors. The prerequisite knowledge of the basic materials science of these systems includes fully understanding principles of synthesis, surface chemistry/derivatisation, and incorporation into macro systems; as this understanding is developed, it should be possible to design microreactors for specific applications. An example of this might be use of nanoscale sensors for direct monitoring of environmental quality as well as associated process monitoring to assure optimum efficiency and minimise energy consumption of equipment. Direct environmental sensors will be important to almost all production process lines, as well as consumer use of vehicles and heating and cooling systems. As we move to smarter and smarter appliances, the need for active sensing and feedback becomes increasingly important to facilitate the "greenest" performance. In many cases, sensing can only be accomplished at the nanoscale.

Synthesis and stabilisation of small nanostructures from a materials science point of view is of environmental concern. It has been demonstrated that very complex crystalline nanoparticles can be synthesised and stabilised by novel chemical and physical deposition approaches. The ability to prepare stable, isolatable particles provides the first major building block towards entirely new important technological areas: the use of nanostructured materials both as precursors to conventional and new materials and as discrete entities in larger passive and active arrays.

Achievements and paradigm shifts

At a time when amorphous metal oxide supports for catalysts were well established within the chemical industry, visionary researchers within Mobil Oil Corporation undertook a long-term research programme into the use of crystalline materials as catalyst supports. That programme has revolutionised catalysis, displacing conventional catalyst supports for many applications In particular, the programme focused on zeolites, porous materials with well-defined shapes, surface chemistry, and pore sizes smaller than 1 nm. The zeolites Y and ZSM-5 are now used widely around the world in several major catalytic processes in the petroleum and petrochemical industries, generating billions of dollars in additional revenues.

Nanoparticles-iron oxides, clays, and other colloids- are the major transporting agents of both pollutants and nutrients in soil, water, and air. Understanding these modes of transport can help distribute desirable organic and inorganic constituents and immobilise undesirable ones. Control could be achieved of morphology, agglomeration, and coatings on natural nanoparticles to have the desired enhancement or retardation effect. Examples currently in use or under development are

1) zeolites and other porous soil conditioners, especially for the controlled storage and release of water;
2) use of clay and zeolites as part of radionuclide barriers at the Yucca Mountain proposed nuclear waste repository;
3) controlled release of iron, phosphorus, and other nutrients from fertiliser;
4) additions of aluminosilicates to food as texturising agents (e.g., in non-dairy creamer) and of zeolites to animal feed (claimed to make pigs grow faster);
5) use of zeolites as ion exchangers in water purification and in detergents; and
6) use of silica gel and other nanophase solids as desiccants.

Many of these are currently low-technology products; fine-tuning them may lead to more sophisticated applications, for example, drug delivery, and environmental regeneration systems in spacecraft and other confined spaces. The ability to identify, characterise, and analyse nanoparticles quickly and in detail with respect to composition, trace elements, atomic-level structure, and morphology will lead to better understanding of their role in pollution and to better, more unambiguous, identification of the sources of pollution.

For example, the definition of "asbestos," currently based solely on shape, needs to be modified to take into account the growing body of data indicating that chemically different particles have very different toxicity. Indeed, the mechanisms of toxicity of nanoparticles are still poorly understood, and nanotechnology-based sensors are likely to play a role in developing such understanding. Similarly, nanotechnology will provide sensors for the detection of low levels of contaminants in air and water.

Applications of nanoporous polymers

A completely new class of organic nanoporous polymers with narrow pore-size distribution has been synthesised using cyclodextrins as basic building blocks. These processable nanoporous polymers exhibit superior inclusion forces and molecular transport properties towards organic guest molecules at water-solid interfaces. In fact, the formation constants of polymeric cyclodextrins and organic guest molecules are over 8 orders of magnitude larger than molecular cyclodextrins in water, and yet the process is completely reversible in organic solvents such as ethanol. The significant potential of these results is that hazardous organic contaminates may be reduced to parts-per-trillion levels in water by these polymers.

Nanotechnology in Environment Restoration

Restoring the environment means returning land to what it was-removing what has been added and, where possible, replacing what has been lost. This can be done, by removing pollutants and some of the pressures for ploughing and paving. A more difficult problem, though, is restoring the ecological balance where the changes have been biological. Much of Earth's biological diversity has been a result of biological isolation, of islands, seas, mountains, and continents. This isolation has been breached, and reversing the resulting problems is one of the greatest challenges in healing the biosphere. A central problem in environmental restoration is reversing environmental encroachment. We tend to see land as being gobbled up by housing, because the land where we live generally is. Farming, though, consumes more land, and the variant of farming called "forestry" consumes still more. By rolling back our requirement for farmland, and for wood and paper, nanotechnology can change the balance of forces behind environmental encroachment. This should make it more practical, politically and economically, for people to move toward environmental restoration.

Human meddling with life in the biosphere has caused enormous ecological disruptions. This hasn't involved genetic engineering-by twisting organisms to better serve human purposes, genetic engineering usually leaves them less able to serve their own purposes, less able to survive and reproduce in the wild. The great disruptions have come from a different source: from globe-traveling human beings taking aggressive, well-adapted species from one part of the planet to another, landing them on a distant island or continent to invade an ecosystem with no evolved defenses. This has happened again and again. People have sometimes tried, with a measure of success, to fight fire with fire: to bring in parasitic species and diseases to attack the imported species and keep its growth within some reasonable bounds. Australia's problem with prickly pear was tackled using an insect from Argentina; the rabbits were cut back-with mixed results-using a viral disease called myxomatosis: "rabbit pox."

Protectors of Ecosystem

Biological controls-fighting fire with fire-have advantages: organisms are small, selective, and inexpensive. These advantages will eventually be shared by devices made using molecular manufacturing, which avoid the disadvantages of importing and releasing yet more uncontrollable, breeding, spreading species. Alan Liss spoke of using nanotechnological devices to help restore ecosystems at a chemical level. A similar idea can be applied at a biological level. The challenge-and it is huge-would be to develop insect-size or even microbe-size devices that could serve as selective, mobile, mechanical flys watters or weed pullers. These could do what biological controls do, but would be unable to replicate and spread. Let's call devices of this sort "ecosystem protectors." They could keep aggressive imported species out, saving native species from extinction.

To a human being or an ordinary organism, an ecosystem protector would seem like just one more of the many billions of different kinds of bugs and microbes in the ecosystem-small things going about their own business, with no tendency to bite. They might be detectable, but only if you sorted through a lot of dirt and looked at it through a microscope, because they wouldn't be very common. They would have just one purpose: to notice when they bumped into a member of an imported species on the "not welcome here" list, and then either to eliminate it or to ensure, at least, that it couldn't reproduce.

Natural organisms are often very finicky about which species they attack. These ecosystem protectors could be equally finicky about which species they approach, and then, before attacking, could do a DNA analysis to be sure. It would be simplest to limit each kind of defender to monitoring only one imported species. Each unit of a particular kind of ecosystem-defender device would be identical, built with precision by a special-purpose molecular-manufacturing setup. Each would last for a certain time, then break down. Each kind can be tested in a terrarium, then a greenhouse, then a trial outdoors ecosystem, keeping an eye on their effects at each stage until one gains the confidence for larger scale use. "Larger scale" could still be quite limited, if they aren't designed to travel very far. This built-in obsolescence limits both how long each device can operate and how far it can move: getting control of the structure of matter includes making nanomachines work where they're wanted and not work elsewhere.

The agricultural industry today manufactures and distributes many thousands of tons of poisonous chemicals to be sprayed on the land, typically in an attempt to eliminate one or a few species of insect. Ecosystem protectors could also be used to protect these agricultural monocultures, field by field, with far less harm to the environment than today's methods. They could likewise be used in the special ecosystems of intensive greenhouse agriculture. Unlike chemicals sprayed into the environment, these ecosystem protectors would be precisely limited in time, space, and effect. They neither contaminate the groundwater nor poison bees and ladybugs. In order to weed out imported organisms and bring an ecosystem back to its natural balance, ecosystem protectors would not have to be very common-only common enough for a typical imported organism to encounter one once in a lifetime, before reproducing.

Even so, as the ecosystem protectors wear out and stop working, they would present a small-scale problem of solid-waste disposal. With the exercise of some clever design, all the machinery of ecosystem protectors might be made of reasonably durable yet biodegradable materials or materials no more harmful than bits of grit and humus in the soil. So their remains would be like the shells of diatoms, or bits of lignin from wood, or like peculiar particles of clay or sand. Alternatively, we might develop other mobile nanomachines to find and collect or break down their remains. This strategy starts to look like setting up a parallel ecosystem of mobile machines, a process that could be extended to supplement the natural cleansing processes of nature in many ways. Each step in this direction will require caution, but not paranoia: there need be no toxic chemicals here, no new creatures to spread and run wild.

If the making and monitoring of ecosystem protectors seems a lot of trouble to go to just to weed out nonnative species, consider this example of the environmental destruction such species can cause. Sometime before World War II, a South African species of fire ant was accidentally imported into the United States. Today, infested areas can have up to five hundred of these ants per square foot. The National Audubon Society-a strong opponent of irresponsible use of pesticides-had to resort to spraying its refuge islands near Corpus Christi when they found these ants destroying over half the hatchlings of the brown pelican, an endangered species. In Texas, it's been shown that the new ants are killing off native ant species-reducing biodiversity. The USDA's Sanford Porter states that due to them, "Texas may be in the midst of a genuine biological revolution." The ants are heading west, and have established a beachhead in California. Without ecosystem protectors or something much like them, ecologies around the world will continue to be threatened by unnatural invasions.

Ecosystem defenders can do flyswatting and weedpulling jobs no humans ever could, but there will also be jobs of shaping, planting, and nurturing. The land has been torn by machines guided by hasty hands, almost overnight. It can gradually be restored by patient hands, whether bare, gloved, or guiding machines able to reshape a ravaged mountain without turning the soil. The green wealth that can be brought by nanotechnology has raised high hopes among some environmentalists. Perhaps people will learn to value nature more deeply when they can see it more clearly, with eyes unclouded by grief and guilt.

Environmental Impact of Molecular Nanotechnology

Molecular nanotechnology could be developed in any of several basically different ways. Each of these basic alternatives itself includes further alternatives. Throughout history, people have worked to achieve better control of matter, to convince atoms to do what we want them to do. This has gone on since before people learned that atoms exist, and has accelerated ever since. Although different industries use different materials and different tools and methods, the basic aim is always the same. They seek to make better things, and make them more consistently, and that means better control of the structure of matter. From this perspective, nanotechnology is just the next, natural step in a progression that has been under way for millennia.

Consider the compact discs now replacing older stereo records: both the old and the new technologies stamp patterns into plastic, but for CDs, the bumps on the stamping surface are only about 130 by 600 nanometers in size, versus 100,000 nanometers or so for the width of the groove on an old-style record. Or look at a personal computer. John Foster, a physicist at IBM's Almaden Research Center, points to a hard disk and says that "inside that box are a bunch of whirring disks, and every one of those disks has got a metal layer where the information is stored.

The last thing on top of the metal layer is a monolayer that's the lubricant between the disk and the head that flies over it. The monolayer is not fifteen angstroms [15 angstroms = 1.5 nanometers] and it's not three, because fifteen won't work and neither will three. So it has to be ten plus or minus a few angstroms. This is definitely working in the nanometer regime. The transistors on computer chips are heading down in size on an exponential curve. Foster's colleague at IBM, Patrick Arnett, expects the trend to continue: "If you stay on that curve, then you end up at the atomic scale at 2020 or so. That's the nature of technology now. You expect to follow that curve as far as you can go." The trend is clear, and at least some of the results can be foreseen, but the precise path and timetable for the development of nanotechnology is unpredictable. This unpredictability goes to the heart of important questions: "How will this technology be developed? Who will do it? Where? When? In ten years? Fifty? A hundred? Will this happen in my lifetime?" The answers will depend on what people do with their time and resources, which in turn will depend on what goals they think are most promising. Human attitudes, understanding, and goals will make all the difference.

Research Directions

Decisions about research directions are central. Researchers are already pouring effort into chemical synthesis, molecular engineering, and related fields. The same amount of effort could produce more impressive results in molecular nanotechnology if a fraction of it were differently directed. The research funders-corporate executives, and decision makers in science funding agencies like the National Science Foundation in the United States and Japan's Ministry of International Trade and Industry-all have a large influence on research directions, but so do the researchers working in the labs. They submit proposals to potential funders, so their opinions also shape what happens. Where public money is involved, politicians' impressions of public opinion can have a huge influence, and public opinion depends on what all of us think and say.

Still, researchers play a central role. They tend to work on what they think is interesting, which depends on what they think is possible, which depends on the tools they have or-among the most creative researchers-on the tools they can see how to make. New tools encourage new thoughts and enable new achievements, and decisions about tool development will pace advances in nanotechnology. To understand the challenges ahead, we need to take a look at ideas about the tools that will be needed.

Throughout history, limited tools have limited achievement. Leonardo da Vinci's sixteenth century chain drives and ball bearings were theoretically workable, yet never worked in their inventor's lifetime. Charles Babbage's nineteenth century mechanical computer suffered the same fate. The problem? Both inventors needed precisely machined parts that were beyond the manufacturing technology of their times. Physicist David Miller recounts how a sophisticated integrated circuit design project at TRW hit similar limits in the early 1980s: "It all came down to whether a German company could cool their glass lenses slowly enough to give us the accuracy we needed. They couldn't."

In the molecular world, tool development again paces progress, and new tools can bring breathtaking advances. Mark Pearson, director of molecular biology for Du Pont, has observed this in action: "When I was a graduate student back in the 1950s, it was a multiyear problem to determine the molecular structure of a single protein. We used to say, 'one protein, one career.' Yet now the time has shrunk from a career to a decade to a year-and in optimal cases to a few months." Protein structures can be mapped atom by atom by studying X-ray reflections from layers in protein crystals. Pearson observes that "Characterising a protein was a career-long endeavor in part because it was so difficult to get crystals, and just getting the material was a big constraint.

With new technologies, we can get our hands on the material now-that may sound mundane, but it's a great advance. To the people in the field, it makes all the difference in the world." Improved tools for making and studying proteins are of special importance because proteins are promising building blocks for first-generation molecular machines. Nobel Prizes are more often awarded for discoveries than for the tools that made them possible. If the goal is to spur scientific progress, this is a shame. This pattern of reward extends throughout science, leading to a chronic underinvestment in developing new tools. Philip Abelson, an editor of the journal Science, points out that the United States suffers from "a lack of support for development of new instrumentation. At one time, we had a virtual monopoly in pioneering advances in instrumentation. Now practically no federal funds are available to universities for the purpose." It's easier and less risky to squeeze one more piece of data out of an existing tool than to pioneer the development of a new one, and it takes less imagination.

But new tools emerge anyway, often from sources in other fields. The study of protein crystals, for example, can benefit from new X-ray sources developed by physicists, and techniques from chemistry can help make new proteins. Because they can't anticipate tools resulting from innovations in other fields, scientists and engineers are often too pessimistic about what can be achieved in their own fields. Nanotechnology will join several fields, and yield tools useful in many others.

Building Small-scale Devices

Today's tools for making small-scale structures are of two kinds: molecular-processing tools and bulk-processing tools. For decades, chemists and molecular biologists have been using better and better molecular-processing tools to make and manipulate precise, molecular structures. These tools are of obvious use. Physicists have recently developed tools that can also manipulate molecules. Combined with techniques from chemistry and molecular biology, these physicist's tools promise great advances.

Microtechnologists have applied chip-making techniques to the manufacture of microscopic machines. These technologies-the main approach to miniaturisation in recent decades-can play at most a supporting role in the development of nanotechnology. Despite appearances, it seems that microtechnology cannot be refined into nanotechnology. For many years, it was conventional to assume that the road to very small devices led through smaller and smaller devices: a top-down path. On this path, progress is measured by miniaturisation: How small a transistor can we build? How small a motor? How thin a line can we draw on the surface of a crystal? Miniaturisation focuses on scale and has paid off well, spawning industries ranging from watchmaking to microelectronics.

Micromachining is also being pursued successfully in Japan and Germany. These microgears and micromotors are, however, enormous by nanotechnological standards: a typical device is measured in tens of micrometers, billions of times the volume of comparable nanogears and nanomotors. In size, confusing microtechnology with molecular nanotechnology is like confusing an elephant with a ladybug. The differences run deeper, though. Microtechnology dumps atoms on surfaces and digs them away again in bulk, with no regard for which atom goes where. Its methods are inherently crude.

Molecular nanotechnology, in contrast, positions each atom with care. As Bill DeGrado, a protein chemist at Du Pont, says, "The essence of nanotechnology is that people have worked for years making things smaller and smaller until we're approaching molecular dimensions. At that point, one can't make smaller things except by starting with molecules and building them up into assemblies." The difference is basic: In microtechnology, the challenge is to build smaller; in nanotechnology, the challenge is to build bigger-we can already make small molecules.

Tools Used for Molecular Engineering

Almost by definition, the path to molecular nanotechnology must lead through molecular engineering. Working in different disciplines, driven by different goals, researchers are making progress in this field. Chemists are developing techniques able to build precise molecular structures of sorts never before seen. Biochemists are learning to build structures of familiar kinds, such as proteins, to make new molecular objects. In a visible sense, most of the tools used by chemists and biochemists are rather unimpressive. They work on countertops cluttered with dishes, bottles, tubes, and the like, mixing, stirring, heating, and pouring liquids-in biochemistry, the liquid is usually water with a trace of material dissolved in it.

Periodically, a bit of liquid is put into a larger machine and a strip of paper comes out with a graph printed on it. As one might guess from this description, research in the molecular sciences is usually much less expensive than research in high-energy physics (with its multibillion-dollar particle accelerators) or research in space (with its multibillion-dollar spacecraft). Chemistry has been called "small science," and not because of the size of the molecules.

Chemists and biochemists advance their field chiefly by developing new molecules that can serve as tools, helping to build or study other molecules. Further advances come from new instrumentation, new ways to examine molecules and determine their structures and behaviors. Yet more advances come from new software tools, new computer-based techniques for predicting how a molecule with a particular structure will behave. Of these fields, it is biomolecular science that is most obviously developing tools that can build nanotechnology, because biomolecules already form molecular machines, including devices resembling crude assemblers. This path is easiest to picture, and can surely work, yet there is no guarantee that it will be fastest: research groups following another path may well win. Each of these paths is being pursued worldwide, and on each, progress is accelerating. Physicists have recently contributed new tools of great promise for molecular engineering. These are the proximal probes, including the scanning tunneling microscope (STM) and the atomic force microscope (AFM). A proximal-probe device places a sharp tip in proximity to a surface and uses it to probe (and sometimes modify) the surface and any molecules that may be stuck to it.

The scanning tunneling microscope (STM, on the left) images surfaces well enough to show individual atoms, sensing surface contours by monitoring the current jumping the gap between tip and surface. The atomic force microscope senses surface contours by mechanical contact, drawing a tip over the surface and optically sensing its motion as it passes over single-atom bumps. An STM brings a sharp, electrically conducting needle up to an electrically-conducting surface, almost touching it. The needle and surface are electrically connected, so that a current will flow if they touch, like closing a switch. But at just what point do soft, fuzzy atoms "touch"? It turns out that a detectable current flows when just two atoms are in tenuous contact-fuzzy fringes barely overlapping-one on the surface and one on the tip of the needle. By delicately maneuvering the needle around over the surface, keeping the current flowing at a tiny, constant rate, the STM can map shape of the surface with great precision. Indeed, to keep the current constant, the needle has to go up and down as it passes over individual atoms.

The STM was invented by Gerd Binnig and Heinrich Rohrer, research physicists studying surface phenomena at IBM's research labs in Zurich, Switzerland. After working through the 1970s, Rohrer and Binnig submitted their first patent disclosure on an STM in mid-1979. In 1982, they produced images of a silicon surface, showing individual atoms. Ironically, the importance of their work was not immediately recognised: Rohrer and Binnig's first scientific paper on the new tool was rejected for publication on the grounds that it was "not interesting enough." Today, STM conferences draw interested researchers by the hundreds from around the world.

In 1986-quite promptly as these things go-Binnig and Rohrer were awarded a Nobel Prize. The Swedish Academy explained its reasoning: "The scanning tunneling microscope is completely new and we have so far seen only the beginning of its development. It is, however, clear that entirely new fields are opening up for the study of matter." STMs are no longer exotic: Digital Instruments of Santa Barbara, California, sells its system by mail with an tomic-resolution-or-your-money-back guarantee. Within three years of their commercial introduction, hundreds of STMs had been purchased.

The related atomic force microscope is even simpler in concept: A sharp probe is dragged over the surface, pressed down gently by a straight spring. The instrument senses motions in the spring, and the spring moves up and down whenever the tip is dragged over an atom on the surface. The tip "feels" the surface just like a fingertip in the simulated molecular world. The AFM was invented by Binnig, Quate, and Gerber at Stanford University and IBM San Jose in 1985. After the success of the STM, the importance of the AFM was immediately recognised. Among other advantages, it works with nonconducting materials. To those thinking in terms of nanotechnology, STMs immediately looked promising not only for seeing atoms and molecules but for manipulating them. This idea soon became widespread among physicists.

Proximal Probes

In 1987, R. S. Becker, J. A. Golovchenko, and B. S. Swartzentruber at AT&T Bell Laboratories announced that they had used an STM to deposit small blobs on a germanium surface. Each blob was thought to consist of one or a few germanium atoms. Shortly thereafter, IBM Almaden researchers John Foster, Jane Frommer, and Patrick Arnett achieved a milestone in STM-based molecular manipulation. Of this team, Foster and Arnett attended the First Foresight Conference on Nanotechnology, where they told us the motivations behind their work.

Proximal probes have advantages as a tool for developing nanotechnology, but also weaknesses. Today, the working tips are rough and irregular. Proximal-probe instruments may be a big help in building the first generation of nanomachines, but they have a basic limit: Each instrument is huge on a molecular scale, and each could bond only one molecular piece at a time. To make anything large-say, large enough to see with the naked eye-would take an absurdly long time. A device of this sort could add one piece per second, but even a pinhead contains more atoms than the number of seconds since the formation of Earth.

Rabbits and dandelions contain structures put together one molecular piece at a time, yet they grow and reproduce quickly. How? They build in parallel, with many billions of molecular machines working at once. To gain the benefits of such enormous parallelism, researchers can either 1) use proximal probes to build a better, next-generation technology, or 2) use a different approach from the start. The techniques of chemistry and biomolecular engineering already have enormous parallelism, and already build precise molecular structures. Their methods, however, are less direct than the still hypothetical proximal probe-based molecule-positioners. They use molecular building blocks shaped to fit together spontaneously, in a process of self-assembly.

A huge technology base for molecular construction already exists. Tools originally developed by biochemists and biotechnologists to deal with molecular machines found in nature can be redirected to make new molecular machines. The expertise built up by chemists in more than a century of steady progress will be crucial in molecular design and construction. Both disciplines routinely handle molecules by the billions and get them to form patterns by self-assembly. Biochemists, in particular, can begin by copying designs from nature. Molecular building-block strategies could work together with proximal probe strategies, or could replace them, jumping directly to the construction of large numbers of molecular machines. Either way, protein molecules are likely to play a central role, as they do in nature.

Protein Engineering

Proteins can self assemble into working molecular machines, objects that do something, such as cutting and splicing other molecules or making muscles contract. They also join with other molecules to form huge assemblies like the ribosome. Ribosomes-programmable machines for manufacturing proteins-are nature's closest approach to a molecular assembler. The genetic-engineering industry is chiefly in the business of reprogramming natural nanomachines, the ribosomes, to make new proteins or to make familiar proteins more cheaply. Designing new proteins is termed protein engineering. Since biomolecules already form such complex devices, it's easy to see that advanced protein engineering could be used to build first-generation nanomachines.

Making proteins is easier than designing them. Protein chemists began by studying proteins found in nature, but have only recently moved on to the problem of engineering new ones. These are called de novo proteins, meaning completely new, made from scratch. Designing proteins is difficult because of the way they are constructed.

A characteristic of proteins is that their activities depend on their three-dimensional structures. These activities may range from hormonal action to a function in digestion or in metabolism. Whatever their function, it's always essential to have a definite three-dimensional shape or structure. This three-dimensional structure forms when a chain folds to form a compact molecular object. To get a feel for how tough it is to predict the natural folding of a protein chain, picture a straight piece of cord with hundreds of magnets and sticky knots along its length. In this state, it's easy to make and easy to understand.

Given the correct conditions, a protein chain always folds into one special shape, but that shape is hard to predict from just the straightened structure. Protein designers, though, face the different job of first determining a desired final shape, and then figuring out what linear sequence of amino acids to use to make that shape. Without solving the classic protein-folding problem, they have begun to solve the protein-design problem. The main advantage of proteins is that they are familiar: a lot is known about them, and many tools exist for working with them. Yet proteins have disadvantages as well. Just because this design work is starting with proteins-soft, squishy molecules that are only marginally suitable for nanotechnology-doesn't mean it will stay within those limits.

Chemists, most of whom do not work on proteins, are the traditional experts in building molecular objects. As a group they've been building molecules for over a century, with ever increasing ability and confidence. Their methods are all indirect: They work with billions of atoms at a time-massive parallelism-but without control of the positions of their workpieces. The molecules typically tumble randomly in a liquid or gas, like pieces of a puzzle that may or may not fit together correctly when shaken together in a box. With clever design and planning, most pieces will join properly.

Chemists mix molecules on a huge scale, yet they still achieve precise molecular transformations. Given that they work so indirectly, their achievements are astounding. This is, in part, the result of the enormous amount of work poured into the field for many decades. Thousands of chemists are working on molecular construction in the United States alone; add to that the chemists in Europe, in Japan, and in the rest of the world, and you have a huge community of researchers making great strides.

An engineer would say that chemists are doing construction work, and would be amased that they can accomplish anything without being able to grab parts and put them in place. Chemists, in effect, work with their hands tied behind their backs. Molecular manufacturing can be termed "positional chemistry" or "positional synthesis," and will give chemists the ability to put molecules where they want them in three-dimensional space. Rather than trying to design puzzle pieces that will stick together properly by themselves when shaken together in a box, chemists will then be able to treat molecules more like bricks to be stacked. The basic principles of chemistry will be the same, but strategies for construction will become far simpler.

Without positional control, chemists face a problem something like this: Picture a giant glass barrel full of tiny battery-powered drills, buzzing away in all directions, vibrating around in the barrel. Your goal is to take a piece of wood and put a hole in just one specific spot. If you simply throw it in the barrel, it will be drilled haphazardly in many places. To control the process, you must protect all the places you don't want drilled-perhaps by gluing protective pieces of metal over most of the wood surface. This problem-how to protect one part of a molecule while altering another part-has forced chemists to develop ever-cleverer ploys to build larger and larger molecules.

Chemists can achieve great things, but have focused much of their effort on duplicating molecules found in nature and then making minor variants. As an example, take palytoxin, a molecule found in a Hawaiian coral. It was so difficult to make in the lab that it has been called "the Mount Everest of synthetic chemistry," and its synthesis was hailed as a triumph. Other efforts are poured into making small molecules with unusual bonding, or molecules of remarkable symmetry, like "cubane" and "dodecahedrane".

Chemists regard themselves as natural scientists even when their life's work is the construction of molecules by artificial means. Ordinarily, people who build things are called engineers. And indeed, at the University of Tokyo the Department of Synthetic Chemistry is part of the Faculty of Engineering; its chemists are designing molecular switches for storing computer data. Engineering achievements will require work directed at engineering goals.

Molecular engineers working toward nanotechnology need a set of molecular building blocks for making large, complex structures. Systematic building-block construction was pioneered by Bruce Merrifield, winner of the 1984 Nobel Prize in Chemistry. His approach, known as "solid phase synthesis," or simply "the Merrifield method," is used to synthesise the long chains of amino acids that form proteins. In the Merrifield method, cycles of chemical reaction each add one molecular building block to the end of a chain anchored to a solid support. This happens in parallel to each of trillions of identical chains, building up trillions of molecular objects with a particular sequence of building blocks.

While it's hard to predict how a natural protein chain will fold-they weren't designed to fold predictably-chemists could make building blocks that are larger, more diverse, and more inclined to fold up in a single, obvious, stable pattern. With a set of building blocks like these, and the Merrifield method to string them together, molecular engineers could design and build molecular machines with greater ease. To make a new molecule, both its structure and the procedure to make it must be designed. Compared to gigantic science projects like the Superconducting Supercollider and the Hubble Space Telescope, working with molecules can be done on a shoestring budget. Still, the costs of trying many different procedures add up. To help predict in advance what will work and what won't, designers turn to models.

Computer-based modeling is already playing a key role in molecular engineering. As John Walker has remarked, "Unlike all of the industrial revolutions that preceded it, molecular engineering requires, as an essential component, the ability to design, model, and simulate molecular structures using computers." This has not gone unnoticed in the business community. John Walker's remark was part of a talk on nanotechnology given at Autodesk, a leader in computer-aided design and one of the five largest software firms in the United States.

Manufacturers and architects know that designs for new products and buildings are best done on a computer, by computer-aided design (CAD). The new molecular design software can be called molecular CAD, and in its forefront are researchers such as Jay Ponder of the Yale University Department of Molecular Biophysics and Biochemistry. Ponder explains that "There's a strong link between what molecular designers are doing and what architects do. Michael Ward of Du Pont is designing a set of building blocks for a Tinker Toy set so that you can build larger structures. That's exactly what we're doing with molecular modeling techniques.

— All the design and mechanical engineering principles that apply to building a skyscraper or a bridge apply to molecular architecture as well. If you're building a bridge, you're going to model it and see how many trucks can be on the bridge at the same time without it collapsing, what kind of forces you're going to apply to it, whether it can stand up to an earthquake.

— And the same process goes on in molecular design: You're designing pieces and then analysing the stresses and forces and how they will change and perturb the structure. It's exactly the same as designing and building a building, or analysing the stresses on any macroscale structure.

— The molecular designer has to be creative in the same way that an architect has to be creative in designing a building. When people are looking at the interior of a protein structure and trying to redesign it to create a space that will have a particular function, such as binding to particular molecules, that's like designing a room to use as a dining room-one that will fit certain sizes of tables and certain numbers of guests.

The engineering perspective helps in thinking about where molecular research can lead: "Even though with nanotechnology we're at the nanometer scale, the structures are still big enough that an awful lot of things are classical. Again, it's really like building bridges-very small bridges. And so there are many almost standard mechanical-engineering techniques for architecture and building structures, such as stress analysis, that apply."

Getting to nanotechnology will require the work of experts in differing fields: chemists, who are learning how to make molecular machines; computer scientists, who are building the needed design tools; and perhaps STM and AFM experts, who can provide tools for molecular positioning. To make progress, however, these experts must do more than just work, they must work together. Because nanotechnology is inherently interdisciplinary, countries that draw hard lines between their academic disciplines, as the United States does, will find that their researchers have difficulty communicating and cooperating.

In chemistry today, a half-dozen researchers aided by a few tens of students and technicians is considered a large team. In aerospace engineering, enormous tasks like reaching the Moon or building a new airliner are broken down into tasks that are within the reach of small teams. All these small teams work together, forming a large team that may consist of thousands of engineers aided by many thousands of technicians. If chemistry is to move in the direction of molecular systems engineering, chemists will need to take at least a few steps in this direction.

In engineering, everyone knows that designing a rocket will require skills from many disciplines. Some engineers know structures, others know pumps, combustion, electronics, software, aerodynamics, control theory, and so on and so forth down a long list of disciplines. Engineering managers know how to bring different disciplines together to build systems.

In academic science, interdisciplinary work is productive and praised, but is relatively rare. Scientists don't need to cooperate to have their results fit together: they are all describing different parts of the same thing-nature-so in the long run, their results tend to come together into a single picture. Engineering, however, is different. Because it is more creative (it actually creates complex things), it demands more attention to teamwork. If the finished parts are going to work together, they must be developed by groups that share a common picture of what each part must accomplish. Engineers in different disciplines are forced to communicate; the challenge of management and team-building is to make that communication happen. This will apply to engineering molecular systems as much as it does to engineering computers, cars, aircraft, or factories.

The advance toward nanotechnology steadily continues. Industry must gain ever-better control of matter to stay competitive in the world marketplace. The STM, protein engineering, and much of chemistry are driven by commercial imperatives. Focused efforts would yield faster advances, yet even without a clear focus, advances in this direction have an air of inevitability.

Nanotechnology is a basic technology, and the Japanese recognise it as such. Recent changes at the Tokyo Institute of Technology-Japan's equivalent of MIT-reflect their views of promising directions for future research. For many decades, Tokyo Tech has had two major divisions: a Faculty of Science and a Faculty of Engineering. To these is now being added a Faculty of Bioscience and Biotechnology, to consist of four departments: a Department of Bioscience, a Department of Bioengineering, a Department of Biomolecular Engineering, and what is termed a "Department of Biostructure." Japan's Institute for Physical and Chemical Research (RIKEN) has broad-based interdisciplinary strength. Hiroyuki Sasabe, head of the Frontier Materials Research Programme at RIKEN, notes that the institute has expertise in organic synthesis, protein engineering, and STM technology. Research consortia in Japan are also moving toward nanotechnology.

Molecular nanotechnology will emerge step by step. Major milestones, such as the engineering of proteins and the positioning of individual atoms, have already been passed. To get a sense of the likely pace of developments, we need to look at how various trends fit together. Computer-based molecular-modeling tools are spawning computer-aided design tools. These will grow more capable. The underlying technology base-computer hardware-has for decades been improving in price and performance on a steeply rising curve, which is generally expected to continue for many years. These advances are quite independent of progress in molecular engineering, but they make molecular engineering easier, speeding advances. Computer models of molecular machines are beginning to appear, and these will whet the appetites of researchers.

Progress in engineering molecular machines, whether using proximal probes or self-assembly, will eventually achieve striking successes; the objectives of research in Japan will begin to draw serious attention; understanding of the long-term promise of molecular engineering will become more widespread. Some combination of these developments will eventually lead to a serious, public appraisal of what these technologies can achieve-and then the world of opinion, funding, and research fashion will change. Before, advances will be steady but haphazard; afterward, advances will be driven with the energy that flows into major commercial, military, and medical research programmes, because nanotechnology will be recognised as furthering major commercial, military, and medical goals. The timing of subsequent events depends largely on when this threshold of serious attention is reached.

In making time estimates, people are prone to assume that a large change must take a long time. Most do, but not all. Pocket calculators had a dramatic effect on the slide-rule industry: they replaced it. The speed of this change caught the slide rule moguls by surprise, but the pace of progress in electronics didn't slow down merely to suit their expectations. One can argue that nanotechnology will be developed fast: many countries and companies will be competing to get there first. They will be driven onward both by the immense expected benefits-in many areas, including medicine and the environment-as well as by potential military applications. That is a powerful combination of motives, and competition is a powerful accelerator.

5

Medical Applications

Nanotechnology will one day revolutionize the field it biotechnology by providing smaller, efficient, and biocompatible biomaterials for use within the body. The field of biomaterials has already been ongoing for quite some time. However, most of the challenge lies in finding ways to create them efficiently. The main goals of biotechnology are to replicate bones, tissues, and organs. Additional goals include hearing and vision implants that could restore lost senses. In all of these respects, nanotechnology has something to offer.

In the field of tissue engineering, nanotechnology promises to improve the efficiency of skin production and regeneration. The quality of artificial skin should also experience a significant increase due to the precision of nanotechnology methods. In the long run, it's expected that nanotechnology will allow entire organs to regenerate or replaced with grown organs. While this seems a little far-fetched right now, it could very well be a reality within a decade.

One exciting prospect is the possibility of cell engineering. The average cell is still in the micrometer regime. If we could tailor individual cells with a network of 'scaffolds' on the nanometer scale on its exterior, we may well be able to significantly enhance tissue properties. An instance of research in this area is a method of regenerating damaged arteries.

Implants are a huge emphasis in biotechnology. The cochlear implant has restored hearing for many people who have ear damage. It's hoped that nanotechnology could further reduce the size and cumbersome nature of today's modern hearing implants. No implant exists yet for restoring vision. Some researchers have successfully created small micrometer scale implants that have stimulated the retina of test animals. It's hoped one day that a dense network of nanoscale photoreceptors will be able to replace a large percentage of a blind person's sight. Though no human subjects have ever been tested, it's only a matter of time before they move to human trials.

The last kind of biotechnology that nanotechnology will effect is bone engineering. It's hoped that with nanotechnology that the complex structure of bone and tooth can be better replicated with some substitute. Nanotechnology involves the creation and use of materials and devices at the level of molecules and atoms. As life itself creates and uses molecular materials and devices, nanoscience will provide great insights in life science concepts, such as how molecular materials self-assemble, self-regulate, and self-destroy. Nanomedicine eventually will infiltrate virtually every field of medicine, if not every realm of human endeavor. Nanomedicine may be defined as the monitoring, repair, construction and control of human biological systems at the molecular level, using engineered nanodevices and nanostructures.

The potential impact of nanotechnology stems directly from the spatial and temporal scales being considered: Materials and devices engineered at the nanometer scale imply controlled manipulation of individual constituent molecules and atoms in how they are arranged to form the bulk macroscopic substrate. This, in turn, means that nanoengineered substrates can be designed to exhibit very specific and controlled bulk chemical and physical properties as a result of the control over their molecular synthesis and assembly. For applications to medicine and physiology, these materials and devices can be designed to interact with cells and tissues at a molecular level with a high degree of functional specificity, thus allowing a degree of integration between technology and biological systems not previously attainable. It should be appreciated that nanotechnology is not in itself a single emerging scientific discipline but rather a meeting of traditional sciences such as chemistry, physics, materials science, and biology to bring together the required collective expertise needed to develop these novel technologies.

Different methods for the synthesis of nanoengineered materials and devices can accommodate precursors from solid, liquid, or gas phases and encompass a tremendously varied set of experimental techniques that are beyond the scope of a short introductory review. In general, however, most synthetic methods can be classified into two main approaches: "top down" and "bottom up" approaches and combinations thereof. "Top down" techniques begin with a macroscopic material or group of materials and incorporate smaller-scale details into them. The best known example of a "top down" approach is the photolithography technique used by the semiconductor industry to create integrated circuits by etching patterns in silicon wafers. This process generally starts by covering a piece of silicon with some kind of photoresist, a photochemicalsensitive polymer that hardens upon exposure to laser light of specific wavelengths. Patterns are then "drawn" into the photoresist-coated silicon with a laser so that the untreated photoresist can be washed away and the exposed silicon treated so it can act as an electrode. Aluminium wires are then placed down in the etched patterns after removing the hardened photoresist to yield transistors. A similar approach has been used to develop microscale connected wells in agar using poly (dimethylsiloxane) molds to study neuron-astrocyte communication. Cell cultures are set up where neurons are in one well and astrocytes in an adjacent well connected by a channel that allows the diffusion of soluble factors. This is an example of a lithographic technique applied to cell biology.

Other types of nanolithographic techniques such as dip pen nanolithography and electrostatic atomic force microscope nanolithography, where individual molecules are deposited or moved, respectively, into desired configurations are able to produce true nanoscale features in various materials. "Bottom up" approaches, on the other hand, begin by designing and synthesizing custom-made molecules that have the ability to self-assemble or self-organise into higher order mesoscale and macroscale structures. The challenge is to synthesise molecules that spontaneously self-assemble upon the controlled change of a specific chemical or physical trigger, such as a change in pH, the concentration of a specific solute, or the application of an electric field.

The physical mechanisms that produce self-assembly, that is, the driving forces that push these molecules to self-assemble into organised structures, are due to thermodynamics and competing molecular interactions including hydrophobic/hydrophilic forces, hydrogen bonding, and van der Waals interactions that aim to minimise energy states for different molecular configurations. The trick is to design systems that self-assemble into macroscopic higherorder structures that display desirable chemical and/or physical properties not displayed by the constituent molecules themselves.A related synthetic approach relevant to biological applications are templating or scaffolding techniques which use pre-existing structures to guide the nucleation and growth of a nanostructured material. Good examples of this are bio-mineralisation type applications such as the growth of artificial bone biomimetics. This is achieved by inducing the formation of organoapatite on titanium structures such as foils; meshes; or porous cylinders by the pre-adsorption onto the bare metal of poly-L-lysine or poly- L-glutamic acid, which when in contact with the metal form a polyionic bilayer. It is thought that the capture of tiny embryonic crystals and subsequent nucleation of these seed crystals are what lead to the actual growth of the apatite on the metal.

Other techniques are attempting to mimic the ultra-structure of bone without the requirement of deposition on pre-existing metal structures by inducing the self-assembly and formation of molecular nanofibers interspaced with mineralised hydroxyapatite crystals, structures that are similar in spirit to the collagen ultra-structure of bone. As these technologies develop, it is expected that these advanced materials will not only provide desirable mechanical properties but also incorporate functional cell signalling properties such as neurotrophic support or anti-inflammatory effects. In addition to various orthopedic applications, this may be of interest to the neurosurgeon when spinal fusions are indicated, for example.

It is important to appreciate that the principle scientific and engineering obstacles involved in developing nanotechnology applications stem from the inherent size and time scales being considered, from a few to about a hundred nanometers spatially and from nanosecond to femtosecond time scales if atomic bond oscillations need to be taken into account, for example. Because the molecular building blocks being manipulated are on such a small scale, these substrates have spatial and/or temporal levels of organisation that span several orders of magnitude, with different levels nested within higherorder levels. Therefore, to study and explore these rich and complex systems, highly sophisticated theoretical and experimental tools are required.

In particular, the visualisation, characterisation, and manipulation of materials and devices require sophisticated imaging and quantitative techniques with spatial and temporal resolutions on the order of 10^{-6} and below to the molecular level. In addition, these techniques are critical for understanding the relationship and interface between nanoscopic and mesoscopic/macroscopic scales, a particularly important objective for biological applications. As such, further nanotechnological advances will necessitate parallel progress of these physical characterisation techniques.

Examples of important tools available at the moment include (a) highly focused (i.e., 1-2 μm) synchrotron X-ray sources for X-ray diffraction and related techniques that provide detailed molecular structural information by directly probing the atomic arrangement of atoms; (b) scanning electron and scanning tunnelling microscopy that allow three-dimensional-type topographical views of nanoscale structures; and (c) in situ monitoring techniques that allow the monitoring and evaluation of building block assembly and growth, such as reflection high-energy electron diffraction (RHEED). The principle behind RHEED is a high-energy electron beam that is directed at a sample at a grazing angle and subsequently picked up by detectors as it bounces off the sample.

The atomic structure of the material being investigated disrupts the path of the high-speed electrons, which produces differing patterns of electron streaks on a phosphor screen detector. It isthese patterns of streaks that are then interpreted to yield the atomic structure of the material being investigated. Ultimately, these and other techniques are required not just to probe the structure of the materials themselves but also to study the interface between these materials and the cells and tissues they are designed to interact with because they are the only means we have of "looking" at what is happening at these extremely small scales.

Recent advances in tissue engineering that have occurred at levels larger than the nanoscales described above include microelectromechanical systems (MEMS) and biocompatible electronic devices that have a significant potential for improving the treatment of many disorders. However, despite this potential, these approaches all depend on bulk molecular engineering or chemical manipulation of macromolecular structures, which lack the ability to manipulate the nanoscale structure. Nanoengineered materials and devices designed to interact with cells and tissues or carry out biologically specific functions should offer a much greater degree of integration between technology and physiological systems. In turn, this should eventually translate into novel clinical applications and treatment options.

At present, applied nanotechnology to medicine and physiology is in its infancy, with most of the research at the basic science stage as the field attempts to organise itself. As such, viable clinical applications are still years away, but despite this the breadth and pace of current research is impressive. One example of such an application includes novel drug delivery systems (specifically for the blood brain barrier in some cases) using nanoparticles or highly porous self-assembling bilayer tubule systems. Another class of applications being developed is chemically functionalised dendrimers, highly branched molecules with a "tree-like" branching structure that can be used as molecular building blocks for gene therapy agents or as magnetic resonance imaging (MRI) contrast agents. Dendrimers are attractive molecules for gene therapy applications because they are a nonviral delivery vehicle for DNA (an important clinical consideration), which tends to wrap itself in the dendrimer branches.

Dendrimer-based MRI contrast agents have been shown to produce increased relaxivity, the induced spin relaxation time of surrounding water protons, which translates into higher contrast images. Specialised membranes for the separation of low-weight organic compounds from aqueous solutions have been developed by filling the pores of microfiltration or ultrafiltration membranes (i.e., pore sizes on the order of microns and smaller) with functional polymeric molecules that have an affinity for the compounds to be filtered. This nanomembrane may allow very selective ultrafiltration of physiologic toxic compounds, for example.

In addition, much research is going into biologically inspired functional nanodevices, such as the development of DNA/polymerase chain reaction or protein-based molecular computers, which encode information in the nucleotide sequences of DNA molecules or in the tertiary structures of proteins and carry out computations by following different biochemical reaction pathways. Another area of interest are biomimetic self-assembling molecular motors, such as the flagella of bacteria, or the mechanical forces produced by RNA polymerase during protein transcription.

These molecular motors provide excellent examples of naturally occurring biological self-assembly, since they are composed of different molecular "parts" that self-assemble to produce the functional structure and are important targets to study and emulate to develop synthetic nanomotors that can interact with biology, for example. These applications represent just a few examples of emerging nanotechnology research aimed at medicine and physiology. Developing nanotechnologies are being pursued for general cellular processes such as ubiquitous signalling pathways that may benefit numerous physiological systems, as well as being targeted toward the particular challenges of specific disorders such as diabetes mellitus or arteriosclerosis. Ultimately, every pathophysiological process has a molecular origin, and it is from this basic fact that the tremendous potential of nanotechnology applications to medicine arises.

Technology is dramatically accelerating the discovery of new drug compounds. Continuing advances in nanotechnology will lead to innovative synthetic routes, new processing strategies, and more economical manufacturing. The same or similar processes that have led to the phenomenal increases in computational speed of microprocessors and the increasing density of computer memory will similarly revolutionise the speed with which new compounds are screened for therapeutic potential as new drugs. The pharmaceutical industry projects nearly a tenfold increase in the number of drug compounds that will be evaluated in 2000 compared to 1998, with only a modest miniaturisation of technology. If the trend is similar to that of microelectronics, the rate could grow exponentially. Arrays of nanodrops, each a mere nanoliter in volume, but holding a small cell culture sample, are being used to place hundreds of thousands of cell culture assays on a laboratory desktop, revolutionising the speed with which new pharmaceuticals can be screened for activity. The time required for new drugs to reach patients could thus be reduced, saving human lives.

Drug and gene delivery will continue to impact significantly on the practice of medicine. Nanotechnology as applied to drug delivery systems will undoubtedly dramatically improve the therapeutic potential of many water-insoluble and unstable drugs. Microsensors interfaced to a nanoscale drug delivery system could dispense precise amounts of drugs for optimum functionality and minimum toxicity. However, significant challenges still remain in synthesis and processing of drug-carrier nanoparticles at the industrial scale. Nanotechnology may also help reach the hitherto elusive goal of active drug targeting to selected cells within the body. Nanotechnology that can further reduce the size and reproducibly attach targeting ligands to the drugloaded nanoparticles may help localise the drug to the desired tissues in the body. These nanoparticles may also be valuable tools for molecular and cell biologists to study fundamental cellular processes such as receptor-mediated endocytosis and intracellular trafficking.

Clinical Diagnostics

The ongoing miniaturisation of electrical components in the semiconductor industry has led to an enormous price-reduction, an increased performance and a strong dissemination of integrated circuits. The ability to control and to guide extremely small quantities of liquids may evoke a similar progress in biochemical analysis and clinical diagnosis. Operating with tiny volumes results in an increased processing speed, reduced costs and thus a higher efficiency, especially when handling rare and expensive materials such as DNA samples. In this regard, microfluidic devices or biochips with analytical and functional capabilities and high sample throughput are receiving increased attention. Two different basic approaches can be distinguished for this "miniaturised" biochemical analysis. The first is based on a parallel processing technique, where a huge number of miniaturised test tubes or wells are placed side by side on a plate. These plates or arrays are commonly named microarrays, for spot sizes below 200 micrometres. Microarrays can be used for diagnostic purposes such as DNA analysis (DNA array, DNA chip or genetic chip) and protein detection (protein array or chip). They offer a high sample throughput due to the simultaneous sample detection, but are limited in their functionality to a single chemical reaction / well. The other approach is based on a serial processing technique, with complex functions integrated on a miniaturised "lab" chip. According to their design, they offer a sequence of different chemical processing and analysis steps, such as mixing of different compounds, synthesis, electrophoreses or chromatography. These systems are called Lab-on-a-chip devices (LOC) and they perform several processes with a minimum material usage. Furthermore, combinations of microarrays and LOC are possible. Besides the common concepts of miniaturised chemistry mentioned above, other microarray systems exist, such as the cell chip or tissue microarrays. These make use of living cells or tissues which are immobilised in discrete spots on an array and permit rapid drug testing or screening experiments.

LOCs may be used for in vitro diagnostics in the future. "Point-of-care" refers to a diagnosis, which is performed on the patient at the bedside in a clinical environment or at the patients home. In the ideal case, the patient would benefit from this near patient diagnosis immediately, as the LOC, which is effectively a functional miniaturised laboratory, would enable the comprehensive analysis of bodily fluids and allow medical staff to determine the optimal therapy. Therefore, point-of-care biochips may help to reduce or even prevent problems arising from drug side-effects and from antibiotic resistance in individuals. In addition LOCs may reduce costs further by removing the need for sample handling and reporting. Taking more than a decade and a tremendous international effort, the decryption of the human genome brought the field of genomics, the study of the functional dependency of genes, to the forefront of science. This rapid development was based on the capability of processing biological data with high throughput and low costs. In this regard, it is believed that biochips will boost the scientific field of genomics in the future. The next step in understanding cellular biochemistry can be found in proteomics, the study of the functionality of all proteins in cells and tissues.

The human genome is the "manual" for synthesising proteins, but it is proteins that provide functions, therefore a comprehensive analysis of disease requires a knowledge of both protein identity and function. Unfortunately, the human proteome is significantly larger than the human genome. A total of one million human proteins has been estimated, whereas the genome consists of about 30000 to 35000 genes. Thus, proteomics is dependent on the availability of high throughput protein-arrays or chips at moderate costs.

Use of DNA Microarray Technology

Arrays and microarrays are used to perform a large number of different experiments simultaneously. They have microscopic and functionalised test tubes or spots, where each spot is designed to perform a distinct test. These reaction spots are arranged in arrays with up to several 100000 different spots on a single chip and a density of up to 100000 spots per cm^2. Microarrays can be used to perform several tasks including DNA sequencing, genetic disease diagnosis, drug efficiency tests (high-throughput screening) and gene expression studies. DNA functionalised microarrays represent one of the most prominent biochips exploiting the correlation of cell function and the expression of genes. Such DNA arrays are available for different genomes including human, mice, rats and micro-organisms, and are commonly referred to as DNA chips. The operation principle of such DNA-chips can be understood by considering the process of events within a cell that converts genetic information into protein, i.e. gene expression.

The information to create a protein, which is stored in a sequence of base pairs in DNA, is copied first into messenger RNA (mRNA), a process called transcription. The mRNA then acts as a template for the synthesis of the particular protein (a process called translation), which fulfils the functional role of the gene. The first stage of gene expression is exploited by gene chips to examine differences between healthy and morbid tissue.

Each spot of a DNA chip contains a molecule of single-stranded DNA. Samples of mRNA to be tested are either labelled (usually with a fluorescent dye) before being applied to the chip or are first converted into cDNA (complementary DNA, i.e. DNA that has been copied from a mRNA) before being labelled and applied to the chip. At this stage, mRNA/cDNA molecules (probe) that match molecules of single-stranded DNA fixed to the chip (target) will bind to these in a process called hybridisation (this is where complementary single-stranded DNA or RNA molecules interact to form a duplex according to base pairing rules i.e. A binds to T, G binds to C), Probes which do not have a complementary, single-stranded DNA target fixed to the chip, will not bind. Thus, only those spots where hybridisation takes place give rise to fluorescence, which can be read by optical means (fluorescence imager). The location and the intensity of fluorescent spots provide information about which genes have been expressed by the cell and to what degree, respectively. In this scenario four different results can occur: i) the red-tagged molecules hybridise, ii) the green-tagged molecules hybridise, iii) both molecules hybridise, iv) no hybridisation takes place. By calculating the relative contribution from each dye, investigators can determine whether a gene from a diseased cell is being expressed to the same, higher or lower levels than in a normal, healthy cell. This allows the investigation of thousands of genes on a single chip and the identification of which genes are involved in specific diseases.

Microarrays are also being exploited for the study of DNA polymorphisms or mutations. Single nucleotide polymorphisms (SNP) are the most common variations in the human genome. These represent a difference of a single nucleotide at a defined point in the genome, for instance when AGGC changes to AGGT. A variation is termed a SNP when it occurs in more than 1 % of the human population. SNPs can be used as a marker in order to identify disease genes and occur on average every 1000 base pairs.

Only 3-5% of the human genome has the potential to encode proteins i.e. are genes. Although most SNPs occur outwith these regions, it is believed that the comprehensive analysis of SNPs within genes will improve medication in the future. If each patient has their own individual SNP map, this could provide detailed information about specific incompatibilities or possible side effects with a given drug. In this regard, a high-throughput method, as can be found in microarrays, is essential. The format of these arrays is similar to that for gene expression studies. A specific oligonucleotide (single-stranded DNA fragment) exhibiting a distinct SNP is immobilised at each reaction spot on the microarray. mRNA is extracted from the patient, introduced to the array where it hybridises to those spots on the microarray with a matching SNP, thus providing information about which SNPs are present in the patient's genome.

DNA microarrays are based on providing a specific "address" for different oligonucleotides. Different single-stranded DNA fragments, representing the variety of genes under investigation, are immobilised on a substrate in discrete spots. Each spot contains a large number of identical molecules increasing the sensitivity. The substrate can be made of silicon-wafer, glass, nylon or other coated materials. Basically, there are two different methods to create the spots containing the oligos. One method is based on the in-situ synthesis of the oligos directly onto the chip surface. The second method takes pre-synthesised oligos and immobilises them onto the chip.

Affymetrix uses the first method to create its GeneChip®. Oligos with a length of up to 25 nucleotides are synthesised using photolithography at each spot of the DNA-chip. The first step involves the immobilisation of a nucleotide at each spot. This, and subsequent, nucleotides have a light-sensitive group protecting the end where further nucleotides are added. In the second step, individual spots are selectively bleached through suitable masks, removing the light-sensitive group and allowing the next nucleotide to be added. By repeating this procedure a tailor-made DNA chip with individual oligos at each spot can be created. Using this method spot sizes of about 20 μm can be obtained, although the costs per chip are comparatively high depending on the number of different oligos synthesised and the number of spots.

Agilent use a second method, similar to inkjet printing, to immobilise molecules inexpensively onto glass chips. The inkjet method deposits the nucleotides sequentially onto treated glass wafers and uses established technologies to elongate these oligos to 60 nucleotides. Spot sizes from 150 to 200 µm can be obtained. However this produces a chip with a considerably lower density than the GeneChip® described above. The ex-situ synthesis methods make use of commercially available printing or spotting methods. These methods can use longer cDNA fragments of up to several thousand nucleotides, but offer a rather limited density.

There are several contact and contact free printing methods, such as contact tip printing, micro contact printing, microsyringe pump printing and piezoelectric printing. The latter two are contact free methods that spray the molecules in small droplets onto the surface. The DNA chip read out process after hybridisation is done predominantly using optical techniques, such as fluorescence microscopy. Alternatives to fluorescent tags include chemiluminescent/luminescent or radioactive markers, and in the future the use of electrical signals is a possibility. However, fluorescent tags have the advantage of multiple colours.

The exploitation of genomics has enormous capability, however proteomics- the study of proteins and their functionality, will be more powerful in the discovery of disease pathways. Despite the progress that has been made in the investigation of the human genome, little is known about proteins involved in pathogenesis. From the genomics perspective this can be attributed in part to the weak correlation between the concentration of mRNA and that of the corresponding protein. Thus, in order to illuminate the function of a given gene in the course of a certain disease, the protein(s) encoded by that gene have to be detected themselves. Considering the complexity and the tremendous number of proteins that are involved in human biology, proteomics seems to be a rather challenging task.

The human proteome is more complex than the genome encoding it would suggest and has been estimated to consist of about 0.5 to 1 million proteins. Unlike the comparatively simple composition of DNA with its four basic building blocks, proteins consist of 20 different amino acids. As a result of interactions between these amino acids, proteins adopt a specific 3-dimensional structure that is a major contribution to their functionality. Proteins can be further modified through the addition of carbohydrate and lipid groups, which adds to the variety issuing from a single gene. Proteins with identical chemical composition but different spatial structures can evoke different reactions. The 3D structure of a protein can be changed when it interacts with other materials including other proteins. Furthermore, proteins are difficult to synthesise and there is no simple way of amplification as can be found in PCR (polymerase chain reaction) for DNA.

At present, the quantitative analysis of proteins is achieved mainly with assistance of two-dimensional polyacrylamide gel electrophoresis (2D-PAGE). Proteins are isolated from cells and initially separated based on charge (first dimension, this is called isoelectric focussing), followed by separation based on mass (second dimension, perpendicular to the first). This method can effectively separate thousands of different protein species on a single gel. Protein "spots" on the gel can be visualised using stains that can either be colourimetric or fluorescent, excised from the gel and analysed by mass spectrometry. Immunoassays, which are based on the specific interaction between an antibody and antigen, heralded the beginning of protein-detecting microarrays. Potential applications for protein chips can be found in diagnostics, drug screening, disease monitoring, drug discovery, and medical research. The market for protein chips is expected to grow to at least 500 million dollars by 2005.

The production of protein microarrays is rather difficult and thus in an early stage of development. Several obstacles have to be overcome in order to produce protein microarrays for reliable high-throughput methods. Firstly proteins exhibit an enormous chemical diversity, ranging from hydrophilic to hydrophobic structures and having huge differences in their size and spatial structure. In order to maintain spatial structure, which is crucial to protein functionality, distinct physical conditions may be required, such as a certain temperature and pH range. Thus, the printing or spotting of proteins has to be achieved without conformational change of the molecules. Furthermore, a specific interaction between the target molecule and the proteins under investigation is required. Unlike the hybridisation of DNA strands, protein-protein interaction is rather complex and driven by their 3D structure. A common approach to protein microarrays is based on antibody arrays, exploiting antigen-antibody interactions.

The basic mode of operation of a protein microarray is similar to that of the DNA microarrays described above. Target molecules are immobilised in discrete spots on the chip surface providing a specific "address" for labelled proteins. The spots where labelled proteins bind targets can be read by optical means (e.g. fluorescence microscopy) providing information about which protein has been expressed. Suitable substrates are glass, silicon or polymer surfaces, which have been chemically treated or coated with appropriate molecules, such as poly-L-lysine, which can bind proteins. The application of the target molecules is performed by a printing method such as inkjet or contact printing. High precision contact printing robots are able to deliver nanoliter quantities of protein samples and to produce spot sizes of 150 to 200 µm and densities of about 1600 spots per square centimetre. Similar to DNA chips, fluorescent or radioactive marking is possible. Although protein chips are still in the initial stage of development, commercial products for protein arrays are available on the market.

Zyomyx Inc. company offers a protein profiling biochip system, which is based on a silicon substrate. The target molecules (antibodies) are immobilised on small three-dimensional pillars for sensitivity reasons. Other approaches rely on polymer coated glass slides, which are covered with a thin (4 µm) HydroGel™ layer (PerkinElmer Inc.). This coating can be used for spotting target proteins in sub-nanoliter quantities using piezoelectric tip technology (Piezoarray, PerkinElmer, Inc.). Since the production of large quantities of proteins is rather difficult, ultra sensitive detection methods are required to measure tiny volumes. Other approaches, besides the common fluorescent and radioactive tagging methods, exist which allow the label-free detection of binding events. One method is based on SPR spectroscopy (surface plasmon resonance spectroscopy), which measures small variations in the surface refractive index of a metallic surface when a protein binds to it. This ultra-sensitive method is capable of detecting the kinetics of biomolecular interaction in real time.

The advantages of microarray technology emanate from their miniaturisation. In this regard, nanoscaled arrays or nanoarrays should have superior properties considering their processing speed and material consumption. Nevertheless, problems arise from the handling of ultra-small quantities of liquids. The reliable processing of volumes down to femtoliters (1×10^{-15} l) is still a challanging task at present and cannot be performed on a routine base. One approach to miniaturise array size can be found in scanning or atomic force microscopy (SFM/AFM).

The AFM is a prominent tool in nanotechnology allowing the visualisation and manipulation of surfaces at the nanometre scale. An atomically sharp tip is used to obtain local topographical information by scanning the area of interest line by line and detecting the forces acting upon the tip through deflection of a microfabricated spring, induced by the scanning process. The AFM can be used as a nanolithographic tool in order to modify the surface structure or chemistry. A common approach is based on the application of a voltage between tip and sample. A different method of AFM nanolithography can be used to produce arrays of nanometre dimensions by the deposition of molecules from the tip through a liquid meniscus (or "ink") onto the substrate during the scanning process.

This technique is called dip pen nanolithography (DPN)and can be used to functionalise surfaces in discrete spots with a user-defined chemistry. The smallest spot size which has been obtained by this method so far is about 100 nm and the ability of DPN to create protein nanoarrays has been demonstrated recently. The inherent problem of all scanning techniques is low throughput which is a result of their serial nature. This can be overcome by intensive parallelisation using several thousands of ink-wetted tips. The advantage of this system can be seen in the label-free detection possibility, which can be done with the same instrument.

Nanotechnology offers advantages not only for array production but also for sample detection on the array. The common approach of fluorescent staining suffers from the disadvantage of bleaching i.e. the fast degradation of fluorescent molecules. In contrast, nanoparticles exhibit improved stability and sensitivity for detection purposes, e.g. resonance light scattering (RLS) of distinct nanosised metal colloidal particles provides an ultra sensitive detection method. A substantial 10-fold enhancement in sensitivity compared to common fluorescent markers has been accomplished through the use of gold and silver particles of uniform dimensions in the range of 40 nm to 120 nm. With regard to protein detection, a new strategy based on functionalised metal nanoparticles, in combination with a magnetic detection method, has been reported recently. This so-called bio-bar code method increases sensitivity and offers a strong multiplexing and simultaneous detection of many proteins.

The current microarray technology is rather macroscopic. Smaller spots can be achieved with scanning probe techniques as mentioned above, however these require a higher detection sensitivity due to the small number of molecules involved. With force spectroscopythe possibility of detecting single molecules by mechanical means can be realised. A conventional AFM with a functionalised tip can be used to measure specific interactions such as antibody/antigen. Typical experiments probe the mechanical properties of single molecules and measure the energy needed to unfold molecular domains.

The miniaturisation of DNA microarrays brought a significant progress in genomics, where the number of experiments covered by a single microarray has been steadily increased. The ultimate gene chip has been produced recently by Affymetrix and has all human genes on a single chip. A better understanding of the functionality of proteins will have a strong impact on future diagnostics and therapies. The development of new drugs allows a more precise targeting with less side-effects. Highly specialised molecules offer an increased effectiveness in the therapy of cancer for example. Such intelligent active agents may require diagnostic tests to determine which patient is able to benefit from the drug e.g. due to the presence of a certain receptor. These tests can be performed using microarray technology in the future, bringing a new approach to diagnosis, the so called personalised medicine, where patients benefit from highly specific developed drugs, which have their effectiveness ensured through a diagnostic test. These molecular diagnostic tools, with precise targeting ability, are referred to as molecular medicine.

Applications of Nanochip Technology

Lab-on-a-chip

The basic idea of lab-on-a-chip (LOC) is the integration of an entire chemistry lab on a small substrate. Resembling the integrated circuits in the semiconductor industry, a LOC can perform sample isolation, mixing, separation, detection and data analysis on one highly integrated platform. In general, a LOC device is able to perform several different production steps, which can vary dependent on the purpose of the LOC, on a miniaturised base. This is achieved by a combination of microfluidic and microelectronic devices creating a small lab, which allows the realisation of a complete experiment with minimum material consumption, and provides data analysis in real-time.

In the future, LOC devices may replace the expensive and time-consuming use of clinical laboratories. Therefore, LOCs are promising candidates for medical analysis at the point-of-care. It was the advent of microfluidics that rendered the miniaturisation of fluid devices possible. Guiding fluids in micrometer-sized channels can be achieved in several ways such as electrohydrodynamic pumping, electroosmotic flow, electrowetting and thermocapillary pumping or temperature gradients. Such an integrated LOC requires micro-engineered surface topographies or a chemical wetting contrast in combination with electronics, which are commonly referred to as micro electro-mechanical systems (MEMS).

LOCs can be fabricated on glass, silicon or polymer substrates, which offers the advantage of a fast and inexpensive production of disposable LOCs. Polymer based microfluidic devices can be manufactured by replication methods such as casting, imprinting or embossing and injection molding, making the mass production of LOCs feasible. The first commercially available product was the Agilent 2100 Bioanalyzer, representing a versatile LOC platform for multiple forms of RNA, DNA and protein analysis. A further miniaturisation of micro-electromechanical systems (MEMS) is leading to the so-called nano-electromechanical systems (NEMS). Although the creation of nanostructured surfaces can be performed on a routine basis at present, problems arise from the guidance of ultra-small quantities of liquids in the pico- and attoliter range.

Nevertheless, nanotechnology provides several approaches for LOC systems. The analysis of nanoscaled objects, such as proteins or viruses, is commonly done by dielectrophoresis. Dielectrophoresis (DEP) uses non-uniform alternating electrical fields to separate and to guide small objects through field gradients. This manipulation requires high electrical field strength, which can be obtained using nanosized electrodes with feature sizes well below 100 nm. Such DEP devices can be implemented in microfluidic systems for the detection of viruses for example.

Nanopore DNA sequencing is another example of a nanotechnology that can be integrated in a LOC. A nanopore, with a diameter of several nanometres, is created in a membrane, which separates two liquid compartments. Single-stranded DNA is passed through the nanopore and as it does it causes variations in electrical current that differ and are specific to each of the four bases of DNA, thus allowing the DNA sequence to be read accurately. LOC devices could serve as point of care devices that would remove the need for the expensive analytical setup of a whole laboratory. In the ideal case, the LOC would be the only instrument needed for diagnosis and it could be carried out near the patient and in real time. However, the common detection method is based on fluorescent markers linked to the sample material. Therefore a fluorescent microscope is still needed to perform the analysis. A different approach may be found in electrical read-out systems, based on standard CMOS technology that is widely used in the semiconductor industry.

Cell Chips

The cell chip is a comparatively new type of biochip, where entire living cells are immobilised on an array. On cell chips protein interactions can be studied without the problem of protein denaturation, which can occur after processing proteins for the use on protein arrays. Therefore, cell chips can be considered as an alternative to protein arrays, which allow the study of the sensitive membrane proteins. On the other hand, cell chips require an enormous effort in their production due to the fact that cells cannot be spotted or printed on the target surface. One approach to cell chip production is based on the functionalisation of discrete spots, providing optimal conditions for cell growth. Cell chips can be used for a number of studies including drug screening, cell membrane ion channels and electrical signals in neurones. A pioneer in the field of cell chips is the company Cellomics Inc., Pittsburgh. They are developing a cell chip with a fluorescent detection system, which will be suitable for drug discovery and screening experiments. For the analysis of the neuronal electrical activity, living neurones have been connected to a sensor array, developed by Infineon and Fromherz et al.. This CMOS based array consists of 128x128 sensors per mm^2.

Nanotechnology is not impacting on cell chips at the present time. Biological cells have a diameter of at least 10 μm and it is impractical to miniaturise the array to these dimensions. Nevertheless, the functionality of the cell membrane is based on nanoscaled features. For example studies of ion channels in the cell membrane involve tapered glass micropipettes with apertures of between several 10 nm and 1.5 μm and a resolution in the same range. The patch clamp technique is used to measure the ion current through the membrane pore. This technique is referred to as scanning ion conductance microscopy (SICM).

Cell chips are at a very early stage of research making it difficult to forecast their future development. They may provide a detailed and comprehensive analysis of specific cell interactions, such as antibody-antigen, and are promising candidates for drug screening and drug discovery. Issues that must still be addressed are the sourcing of cells (ensuring that defined, homogeneous populations of cells are used) and reproducibility.

Use of Imaging Techniques

Imaging techniques are widely used in medicine and biochemical research. Techniques such as x-ray, computer tomography (CT), ultrasound (US), magnetic resonance imaging (MRI) and nuclear medicine (NM) are well established for the diagnosis of disease without surgery. The information derived from each method is complementary, so that often several methods are needed for a full diagnosis.These methods are used mainly to analyse anatomy and morphology and not changes at the molecular level.

Nanotechnology has no great influence on this traditional use of imaging techniques, but new developments in biochemistry and molecular biology extend some of the above-mentioned methods to a higher sensitivity and specificity at the molecular level. This falls under the banner of molecular medicine whose aim is to detect the early biological markers of disease and diagnose illnesses before the onset of the first symptoms. The field of molecular imaging is part of a paradigm shift to the maintenance of health instead of the treatment of disease. Information on the molecular level is of great importance to an early diagnosis and a more accurate characterisation of disease, as well as early indicators of the success of treatment.

A 'payload' of either a diagnostic 'signal giver' and/or a therapeutic agent is tagged to a delivery agent. The delivery agent can be a small molecule, or artificially created carriers such as nanoparticles or microbubbles. The targeting system refers to a specific binding process or molecule, characteristic of the biochemical process to be imaged and/or treated. This basic concept is not restricted to nuclear medicine. It can also be applied to MRI, ultrasound and optical methods.

Nuclear imaging techniques require the injection of a radioactive substance, which emits gamma rays. The absorption of radionuclides in tissues is dependent on metabolism and allows information about physiological processes, anatomy and organ function to be gathered. The main application areas are oncology and cardiology. However the disadvantage of nuclear imaging is that it delivers limited morphological information, which makes it difficult to precisely localise regions with higher metabolism. A special method of nuclear imaging is positron emission tomography (PET). By using two opposing detectors the sensitivity is increased strongly which makes PET a promising method for molecular imaging.

Magnetic Resonance Imaging (MRI)

Magnetic resonance imaging (MRI) scanners, require strong magnets to align the nuclear spins of hydrogen atoms in tissue in one direction. High frequency coils excite the nuclear spins and analyse the weak signals of their relaxation process, with gradient coils used for further, local resolution. MRI is known for its ability to generate high-resolution images, displaying fine anatomical details at sub-millimeter dimensions. MRI scanning is generally regarded as being harmless to the body, and can be applied repeatedly, with almost no restrictions, making it the ideal modality for the evaluation of treatment. Its limitations arise from similar proton densities in different tissues. Therefore, various contrast agents are used, which are concentrated in certain tissues. Smart design of the contrast agents allow molecular signatures to be visualised down to picomole levels. The 2003 Nobel prize in physiology or medicine was awarded to Paul C. Lauterbur (U.S.) and Sir Peter Mansfield (UK) for their discoveries concerning MRI.

Ultrasound Diagnostics

In ultrasound diagnostics, the intensity of the ultrasound wave is reflected at tissue or organ boundaries. The reflected intensity is measured and displayed as an image on a monitor. The contrast between healthy and diseased tissues is usually weak. However, over the last two decades, several contrast agents have been developed, along with contrast-specific imaging methods, to enhance their discrimination. Current research is looking for contrast agents that bind to molecular markers. Traditional contrast agents consist of microscopic gas bubbles (microbubbles) surrounded by a stabilising coating to protect against dissolution. This well-established method offers a high sensitivity-it is even possible to detect a single microbubble. When the microbubbles are injected intravenously, they act as blood pool agents, which limits their application to targets within the vascular system.

Optical Imaging

Optical imaging is an important tool in life sciences for the detection of gene expression and protein-ligand interactions. Although many of the techniques are restricted to in vitro applications due to problems with optical access or labelling, optical imaging is being used increasingly for in vivo imaging as well. Absorption, reflectance, fluorescence, or bioluminescence can be used as the source of contrast.

Currently, most of these techniques are primarily restricted to microscopic or surface imaging, or to experimental imaging in small animals, since the penetration depth of the light is limited. However, light within a small spectral window of the near infrared region (600-900 nm) can penetrate more than 10 cm into tissue, due to the relatively low absorption coefficient at these wavelengths. The lowest wavelength in this window is determined by the relatively high absorption of blood (haemoglobin), whereas absorption above 900 nm is due to water. Near-infrared (NIR) fluorescence and luminescence imaging make use of this optical window.

Usually, optical contrast agents have to be used to obtain high specificity and sensitivity. In fluorescence imaging a fluorescent probe (optical contrast agent) is activated by an external light source, and a signal is emitted at a different wavelength. The fluorescence signal can be captured with a high-sensitivity charge-coupled-device (CCD) camera. The sensitivity of fluorescence detection is very high, and in microscopic setups it is possible to detect single molecules. Although in vivo optical imaging is a powerful tool in cell culture studies or animal models, its current application to human health is generally limited to 'close-to-surface' structures, as in optical imaging of the eye and skin, or optical mammography. In addition, the widely used dyes, called fluorophores, have drawbacks. They fade quickly, and only two or three colours can be used simultaneously to label different cellular components.

Electron Microscopy

Pathology represents a completely different field of diagnosis, e.g. where analytical methods are used to search for viruses in biopsy samples. Viruses have sizes of around 100 nm or less and therefore their detection is suited instruments such as the electron microscope (EM), which is used for analyses in the nanometre region. For example, polyomaviruses (PV) are pathogens which have tropism to the human urinary tract and have been reported with increasing frequency in renal transplant recipients.It is difficult to distinguish the effect of PV on graft rejection from patient immune responses against transplanted renal antigens.

An accurate diagnosis is of great importance because the therapeutic approaches to both are very different. PV appear as 40 to 50-nm, non-enveloped particles within epithelial cell nuclei and can be readily identified using EM. The EM was also the "first line of defence" during the SARS (Severe Acute Respiratory Syndrome) breakout. SARS is a viral respiratory illness that was first reported in Asia in November 2002. It is caused by a coronavirus, called SARS-associated coronavirus (SARS-CoV). The virus has a diameter of about 100 nanometers and was identified through an electron microscopic analysis.

Use of Naotechnologies in Drug Discovery

Drug discovery requires an identified disease, a knowledge of the disease mechanism, thereby identifying a point of intervention, and a model system to test efficacy.Most drug targets are proteins. These can be human proteins that are either defective or abnormally expressed, or can be proteins within a pathogen. The site of interaction of the drug with target is often at the protein's active site. This can be likened to the functional part of the protein, e.g. for DNA polymerases this is where bonds between nucleotides are catalysed to make DNA.

Novel drugs have been discovered by a variety of different routes ranging from serendipity (e.g. penicillin), the screening of natural products to combinatorial chemistry where chemicals are systematically modified to produce large numbers of variants and then screened against targets. By understanding the process involved in a disease and approaching this rationally, compounds with known beneficial effects can be chemically modified to make them more effective, safer or more easily administered. Mass screening of chemicals against targets can be used to identify which ones have the desired effect.

Traditionally the pharmaceutical industry followed an empirical route to drug discovery. By understanding the mechanism of a disease compounds could be synthesised to evaluate their ability to interact with targets and alleviate disease symptoms. This approach relies on a one drug one target cure, with pharmaceutical companies hoping to produce a so-called "blockbuster drug", each of which can generate in excess of $1 billion per annum.

However this has identified fewer and fewer new drugs or new molecular entities (NMEs) over the last 10 years, and of those that have been produced the majority fall into the "me too" category. Such drugs are novel formulations against established targets for which another company already markets an effective drug.

With the sequencing of the human genome novel approaches to drug discovery have become evident. The human genome contains approximately 35,000 genes each of which can be copied into messenger RNA (mRNA), which acts as a template for the translation of the genetic code into proteins, which in turn exert the biological (or phenotypical) effects of the gene. However for more than 50% of human genes this biological effect has yet to be identified.

The complexity of the human genome is raised further through tissue- and developmental-specific expression of genes, the potential for some genes to express multiple forms of mRNA and through modifications of translated proteins, all of which leads to the differential expression of proteins and forms of the same protein (isoforms) within different tissues and even the same tissue under different physiological conditions. All-in-all this increases the total number of encoded proteins to more than 100000, and at present, potential targets for drug intervention have been estimated at anything between 600 and 10000.

Genomics looks at all the genetic information in an organism and feeds into an investigation of the transcriptome (or the total complement of mRNA in a cell under the studied conditions) and ultimately proteomics, which is the study of the total cellular protein complement. The technology allows an investigation and a comparison of each of these in terms of mutations at the DNA level, and differential mRNA or protein expression. Nanotechnology is becoming key to the success of these endeavours through the miniaturisation, automation, speed and reliability of assays.

Target Identification

The first stage of drug discovery is the identification of a target that can be subsequently acted on by a future drug to bring about an improvement in disease. This can be achieved more quickly and reliably now through the use of array technologies.Up to 500000 oligos per square cm can be packed onto slides, allowing the detection of thousands of genes or mRNA species on a single chip. Differences in gene expression between normal and diseased cells can be determined by incubating microarrays with cDNA (mRNA which has been copied back into DNA) isolated from both sources that have been labelled with different fluorescent tags. As each spot is associated with an attached oligo of known sequence, the relative fluorescent intensities from each tag at each of these spots allows the determination of the relative expression of different genes in each source. By establishing whether the transcription levels of specific genes are the same, higher or lower in the diseased cell compared with normal, novel targets for drug intervention can be identified.

For example oestrogen has been known to play a key role in the development of breast cancer since it was observed that a side-effect of chemotherapy was inhibition of oestrogen production in the ovaries. Earlier this year Frasor and colleagues used DNA microarrays to identify oestrogen responsive genes in breast cancer cells, and therefore potential targets for drug intervention. According to analysts the market for DNA microarrays was $400-550 million in 2000, and this is expected to grow to $1-2.2 billion by 2005. DNA microarrays provide a robust platform for the screening of multiple candidate drug targets, however as has been mentioned previously the diversity of proteins in any cell is not restricted to the number of genes that it expresses. Therefore to identify all possible targets, and in some cases to identify any targets, it is necessary to look at the protein complement of the cell. Proteomics has been defined as "the branch of genetics that studies the full set of proteins encoded by a genome." Proteomic tools have been traditionally less robust and more time-consuming than those for DNA. However recent advances in protein chips and a fusion of DNA and protein technologies may change all that.

2D-polyacrylamide gel electrophoresis (2D-PAGE) separates proteins based on charge and mass, with it being possible to analyse two populations of proteins (e.g. from healthy and diseased tissue) stained with different fluorescent markers analysed on the same gel. Following electrophoresis, protein "spots" are excised and identified by mass spectrometry. This technology has been bolstered recently through the production of protein chips.

The identification of a protein(s) which is differentially expressed between diseased and normal tissues is the first step. If the protein has been previously characterised and its biological function is known, then perhaps intervening drugs are already available or can provide a starting point for the elaboration of more effective drugs. If however, the protein is unknown then its function must be either determined experimentally through expression and in vitro studies, or through the relatively new field of bioinformatics.

The "fusion" of DNA and protein technologies has come from the development and use of aptamers. Single-stranded DNA and RNA molecules can form complex 3D shapes through intramolecular interactions, and in some cases these formations (or aptamers) can interact specifically with proteins (much in the way that antibodies do). Somalogic have incorporated bromodeoxyuridine into single-stranded nucleic acids to generate photoaptamers which can crosslink to proteins when irradiated. This provides a handle for the detection of specific proteins within a sample. Photoaptamers can be arranged in array format, different populations of proteins can be incubated on matching arrays, proteins UV-crosslinked and differences in individual protein expression levels determined by applying a universal protein fluorescent stain to the plates.

Target Validation

The genomic and proteomic methods described above can identify differences gene expression and protein levels between normal and diseased cells. The use of DNA microarrays in particular has lead to an explosion in the number of candidate targets. But how do you tell whether these candidates are significant? How do you tell whether the target that has been identified will be amenable to intervention and whether it is the sole factor in the disease or is part of a complex pattern of differential protein expression? These questions must be answered by target validation. According to David Szymkowski, director of biotherapeutics at the biopharmaceutical company Xencor in Monrovia, California, drugs fail in the clinic for two basic reasons: they either don't work or they prove to be unsafe. "Both of these are often the direct result of sloppy early target validation." It is important to ensure that the correct target is identified before candidate drugs are tested against it, as it can take 12 years and up to $1 billion dollars to bring a drug to market. Traditionally, target validation was achieved *in vivo*, however recently *in vitro* models have been developed which, although not available for more complex disease syndromes (e.g psychological diseases), offer a more ethical, quicker and cheaper alternative to animal models.

The targets of many of the blockbuster drugs were validated using gene knockouts. Gene knockouts (KO) involve deleting or disrupting expression of a gene in early embryogenesis in an animal model (usually mouse, but more recently zebrafish have been used). The rationale for this is that removing the gene encoding a protein that has been identified as a candidate will effectively mimic the application of a strong repressor for that protein, allowing investigators to determine the protein's normal cellular function. For example KO mice for the histamine H_2 receptor are unable to secrete gastric acid into the stomach in response to histamine, thereby preventing lowering of stomach pH and validating the receptor as a target for the treatment of gastric ulcers. KO animal models have to be taken in context. Some mutations may be lethal, some may have differential effects depending on developmental stage and others may have multiple effects.

Using tissue culture systems is quicker and cheaper than animal models. Genomic approaches include antisense oligonucleotides and RNA interference. The principle of antisense technology is the binding of an oligonucleotide to complementary mRNA in the cell. This can prevent gene expression in one of 3 ways: 1) directed degradation of the DNA-RNA duplex by a cellular enzyme RNase H, 2) blocking protein translation or 3) preventing RNA processing, prior to translation of protein. BIOGNOSTIK manufacture phosphorothioate antisense oligonucleotides (ODN) that are capable of entering cells in tissue-culture and binding to complementary mRNAs preventing their translation. In the last few years RNA interference (RNAi) has received much attention for its ability to silence genes. This technology makes use of a natural cellular defence mechanism against double-stranded RNA (dsRNA) molecules, which can arise during virus replication for example. Short dsRNA molecules that are identical in sequence to the target mRNA are transfected into cells where they are cleaved into smaller fragments by a cellular enzyme called Dicer.

There are many more proteins than genes, and so validating targets at the DNA or mRNA level may not always present an accurate physiological picture. Proteomics approaches are now appearing in the form of solid phase and solution arrays. Xerion Pharmaceuticals use a novel procedure to functionally inactivate target proteins. First the target is "panned" against a phage display library that expresses single chain antibody fragments (scFv) which bind proteins with high specificity. Specific scFv's to the target are identified, enriched and labelled with UV-reactive dyes. These labelled scFv's are allowed to interact with the target protein within the cell and then irradiated using a laser. This causes the breakdown of the dyes into active compounds which in turn cause the inactivation of the target protein. This procedure can also be used for a disease-based approach, combining target identification and validation. Proteins are isolated from healthy and diseased cells and scFv's identified that only bind proteins from diseased cells. These are used to validate the target as described above, and can then be used to isolate the target for identification by mass spectrometry.

Following identification and validation of a target in a disease, drugs must be designed and validated that will affect that target to alleviate disease symptoms. First of all lead compounds must be generated. This can be achieved through modelling or molecular manipulation. These then must be screened to confirm their efficacy and determine which compounds will be developed further into marketable drugs.

Structure Guided Discovery

In the 40 or so years since Perutz and Kendrow determined the 3D structure of haemoglobin and myoglobin, X-ray crystallography has proven invaluable in the generation of data on the 3D structure of proteins. It has become clear that proteins that perform similar functions but have different substrates (e.g. kinases, proteins which phosphorylate and thereby modify the activity of other proteins) have similar structures around their active sites.

More recently, bioinformatics (defined as the merging of biology, computer science, and information technology into a single discipline) has been used to predict the 3D structures of targets and potential chemicals that could interact with them (ligands) and inhibit protein function, by comparing primary amino acid sequence of novel proteins with the determined 3D structure of known proteins. This approach has been heralded as the new way forward in drug discovery. It is estimated that only 1 out of 5000 screened compounds is approved as a new medicine. By using modelling systems many compounds can be eliminated from screening before they are synthesised.

De Novo Pharmaceuticals uses complex algorithms in its Site Explorer™ software to characterise the active site of a target and determine potential structures of ligands which can interact with it. Candidates are then screened "in silico" using the company's EasyDock programme to determine relative binding affinities. Using this approach thousands of compounds can be screened and only the best candidates synthesised for screening. A unique approach is employed by Momenta Pharmaceuticals. 80% of known proteins are modified through the addition of carbohydrate molecules, and it has been known for a long time that these "post-translational" modifications can influence protein function. Momenta Pharmaceuticals have developed a technology which allows the identification and functional characterisation of these sugar groups, which in turn can feed into the development of new therapeutics.

Molecular Manipulation

Traditionally drugs were designed and modified in what could be considered a linear approach. A compound was identified and if for whatever reason it had drawbacks, it was modified in a rational manner based on reactive groups within the molecule. This approach however, can take a long time to generate better compounds. Combinatorial chemistry however can be likened to taking many different LEGO building blocks and putting them together randomly to create many different structures. The building blocks are small reactive chemical species and the larger compounds that they combine to form are candidate drugs which can be generated in their thousands and then assayed on the target for functionality.

Alantos use a procedure termed Target Amplified drug Candidate Evolution (TACE) to generate libraries of compounds that "can evolve via interaction with the biological target to produce preferentially or exclusively only those compounds that bind tightly to the target." Pharmacopeia also generate vast chemical libraries and have designed high-throughput screening assays to test compounds on different target proteins. By generating libraries of molecules, companies are not only maximising their chances of finding an effective drug, they are also going some way to fending off other companies future efforts to produce and market "me too" drugs.

An alternative method is to combine smaller molecules in the presence of the target. This has been successfully used to develop inhibitors against the influenza neuraminidase protein, which is a major target for anti-influenza therapies. Hochgürtel and colleagues mixed specific aldehydes and amines in the presence of neuraminidase, allowed these to chemically react and products, which resulted in enzyme inhibition, were identified by chromatography and mass spectrometry.

Improvements Surgery Field

Surgery has been practised for a long time, with surgical tools found from cultures as long ago as ancient Rome. With simple tools and techniques "surgeons" could provide critical help for injured persons. Of course this often created infection, toxicity and inflammatory problems due to the lack of sterile clinical conditions or of knowledge in how to prepare tools. However tools were developed for almost maximum effect and surprisingly enough some of the designs from back then can be found in present instruments. The parallel research and development of tools and equipment on one side and surgery techniques on the other, has led to massive improvements in the surgery field. Tools are now very well developed and every factor for securing a patient's health is taken care of- including the provision of a sterile environment around and within the patient. Recent developments such as laser surgery have been used for the successful removal of cancerous tissue for example.

Laser surgery uses concentrated beams of light of particular wavelengths instead of instruments for surgical procedures. Lasers were first developed in 1960. Since that time, many modifications have been made and a number of different types of lasers have been developed. These differ in emitted light wavelengths and power ranges, as well as in their ability to clot, cut, or vaporize tissue. The most commonly used are the pulsed dye laser, the Nd: YAG laser, the CO_2 (carbon dioxide) laser, the argon laser, and the KTP laser. Each works in a different manner and each is sometimes used as part of a treatment plan. The laser works as following; when the laser beam hits tissue such as skin, a certain amount of light is reflected. The remaining light is absorbed or transmitted through the tissue. Absorbed light has specific effects, creating physical, mechanical, chemical, or temperature changes that result in the desired surgical effect. Lasers permit highly localised precise surgery with reduced blood loss. Moreover, a relatively dry operative field affords better vision for the surgeon.

In what is generally referred to as open surgery, surgeons have traditionally had to make an incision large enough to expose the internal organs being operated on, i.e. to gain access to the body cavity, normal, healthy tissue is cut. Within the last decade, however, new minimally invasive surgical techniques have revolutionised many surgical procedures by enabling surgeons to access body cavities for procedures while reducing the trauma to overlying skin, muscles, and nerves that occurs with traditional open surgery. Most minimally invasive surgical procedures are performed with small optical instruments, such as endoscopes, and instruments placed through the body. The benefits of minimal invasive surgery are a decreased injury to skin, muscle, and nerves, fewer wound complications, easier postoperative pain management and quicker recovery and return of the patient to normal functioning

All the surgical instruments available on the market today, varying from scalpels and knives to scissors and retractors. Usually these are made of stainless steel or tungsten but more recently titanium instruments have been introduced. Titanium benefits from its light weight and high tensile strength producing a much lighter instrument and giving the surgeon better handling during an operation. Sutures are another essential medical device and have been used since the beginnings of surgery to close up tissues or compress blood vessels and prevent bleeding. Historically hair, animal gut and various textiles were used, which understandably caused many problems for the patient, however these have now been replaced by synthetic materials such as polyethylene and polypropylene.

The driving force nowadays is to make surgery progressively less invasive. By producing a new set of tools on the nanoscale one can perhaps imagine a surgeon being able to make changes to and track individual cells. This could be very beneficial for neurosurgical aspects for example, and, in addition, the patient will benefit from the reduced trauma of even smaller wounds. Gesellschaft für Diamantprodukte (GFD) have, through the application of nanotechnology, in a process termed "plasma polishing", created diamond scalpels with a cutting edge of only a few atoms (approximately 3 nm), that have applications in eye and minimal invasive surgery. The width of the scalpel blade is approximately one thousandth that of a metal blade and makes these scalpels officially the smallest in the world.

Minimally invasive surgery will also answer another demand, not only from surgeons but also patients, which is to make the operation as short and minimally traumatic as possible and reduce the length of time spent in the hospital. Another way to initiate this is to establish virtual surgical setups. The introduction of telepresence surgery has the potential to remove surgical limitations between countries and continents. However political, ethical and cultural differences between countries will need to be addressed to ensure that suitable procedures, acceptable to all parties, are established.

When performing a surgical operation the surgeon works in a biological environment with corrosive or other effects. As a result it is essential that instruments are thoroughly cleaned and rinsed with de-ionised water before sterilisation, to avoid corrosion and pitting. Corrosion of an instrument can lead to its failure (e.g. not opening or closing properly) and can visually mask contamination, which in turn can spread infection to patients who are operated on. The use of titanium alloyed with nano-composites of aluminium and/or vanadium in new surgical instruments addresses this. Unfortunately such alloys are extremely difficult to machine and finish, currently making the prices for these products very high. Titanium is immune to corrosive attacks by laboratory saline solutions, blood and other bodily fluids, chlorides, oxidizing acids, alkalis etc., and it is 100% nonmagnetic, a great benefit when working with nerves, for example.

Not all surgery techniques will be influenced through the production of better tools. For example sodium-2-mercaptoethane sulphonate (or MESNA) has been described recently as a "chemical dissection agent". This compound breaks disulphide bonds between proteins allowing tissue to be pulled apart more easily. As a result MESNA could potentially reduce surgery time, blood loss, trauma to local tissues and post-operative complications.

Considerable efforts are being made in the creation of new tools for minimally invasive surgical instruments. Over the past decade, a number of ingenious instruments have been developed for manipulating individual cells, macromolecules and even individual atoms. One prime example is the scanning probe microscope, developed originally at IBM Zurich Research Labs. These tools have contributed in significant ways to advances in several fields of research, such as electron transport in molecules, protein folding mechanics and dynamic processes at cell membranes. Such research provides novel insight into the structure and function of molecules relevant to biotechnology and the surgery field.

Virtual Reality (VR) interfaces to experiments have also evolved rapidly over the past decade, a notable example being the nanomanipulator at the University of North Carolina, the leading U.S. academic VR research center, which provides a virtual interface to a scanning probe microscope. This device has now been commercialised, and has been successfully used in teaching programmes. Another major development of the past decade has been the use of the Internet and PCs to control scientific equipment and analyse experimental data remotely. A prime example is the World Wide Web, which was originally developed by CERN for scientific purposes. Today, researchers are developing the GRID, a next generation of network applications that will greatly facilitate internet-based distributed computing and remote visualisation, including VR applications. This will be particularly important for surgical operations as it requires a lot of computer power to interpret and display images and data to allow the surgeon to make the correct decisions.

The surgeon uses a haptic display that delivers force feedback from the operation scene. A simple example of how this works is the steering wheel found in some computer games, where a servo motor can make the wheel turn and vibrate just as a real wheel on a real car would do. In the nanomanipulator, the haptic device is a pen-like rod connected to several servos making the pen capable of 3D motion. When connected to a computer with appropriate software, the pen can be used as a probe to "feel" the outline of simulated objects. If the device and the software are advanced enough, texture and elasticity can also be included. Instead of "feeling" a simulated landscape, the haptic device can be connected to an atomic force microscope (AFM). In this case, the AFM tip is moved in the same way as the pen, and the hand guiding the pen feels the surface magnified about a million times. If there are particles lying on the surface, the pen can be used to move them about. Once in a while, the AFM is run in tapping mode to image the surface so that the operator can see what has been created.

Application of Nanotechnology to Implants

The technology behind implants has been achieved through years of research and development. Each year about 430 000 patients in the European Union have a total hip or knee joint replacement to provide relief from crippling joint diseases such as arthritis. However with everyday wear the implant surface degenerates, producing a tremendous number of predominantly sub-micron-sized polyethylene particles, and eventually leading to implant loosening. As a result the survival time of these implants rarely exceeds fifteen years. Implants have been used for a long time, e.g. people have used gold for dental repair for more than 2000 years and the Aztecs used wood to fix broken bones. As early as the beginning of the 20th century other materials such as metals were used for implants.

New developments and discoveries during and after World War II led to experiments with materials that are unfamiliar as present day implant-materials. Materials such as Dacron fabric were used for prosthesis in cardiovascular surgery and Plexiglas was used for artificial eyes. This caused many problems for users as the technology was not sufficiently advanced to minimise toxicity and infection. In addition, research and development of better implants at this time was very much on a trial and error basis. The implants were often made of regular materials on the market and not necessarily developed with the biocompatibility in mind. Later on materials such as titanium, cement and polymers were introduced to address biocompatibility problems. All of these improved the situation, however there are still problems such as wear debris and release of degradation products. One key issue to address when developing successful implants, that remain functionally intact over a long period of time, is how the new implant will interact with the body. This depends on the implant material, its biocompatibility and whether or not the mechanical properties will remain the same over time.

The application of nanotechnology to implants allows the structure or surface of a biomaterial to be precisely defined thus improving biocompatibilty. Nanotechnology can also play a significant role in decreasing toxicity by creating a nano-coating on the implant, which reduces the effect of a corrosive environment. Optimising the coating not only avoids toxicity but also establishes a better interaction with the surrounding cells and tissue leading to a more stable and biocompatible implant. A nano-coated titanium or tantalum implant will therefore be better adapted to the human body and less likely to need to be subsequently removed and replaced by a new implant. It is however important that the coating does not itself cause problems by debonding.

Biocompatibility is the ability of a material to perform with an appropriate host response in a specific application. An implant should be biocompatible in order to stay in the body over time and interact properly with living tissue. Biocompatibility can be achieved in several ways. Either an implant can be made from a biocompatible material and then implanted directly. Otherwise, biocompatibility can be achieved by coating the implant with a biocompatible material, or modifying the surface in another way. Several technologies for surface modification are used, including self-assembled monolayers, lithography, microcontact printing, colloidal lithography and protein template imprinting. In addition, the implant can be chemically treated before use, e.g. for direct bone bonding.

Biomimetics is the technology of biological growth of new bio-structures in a mould or directly in/on the injured bodypart. Researchers develop these structures from a single nano-physical or nano-chemical structure. A nano-physical structure is developed from a single nano-crystal, whereas a nano-chemical structure is developed from an array of large reactive molecules attached to a surface. The idea is to use these nanostructures as seed molecules, or seed crystals, to drive materials to grow by themselves (*in vitro* or moulding). Researchers hope that this will lead to much better biomaterials for new and better implants.

Other trends in the biomimetic field are the understanding of how specific body parts evolve and develop. Researchers are working on constructing replacement tissues such as muscle and bone using "tissue-engineering" techniques. Also the understanding of the biological functions of molecules e.g. biological motor proteins is now given lot of attention. The development of a molecular motor will aid the understanding of the biological process involved in muscle contraction and expansion.

Research in Nanomedicine

There is considerable anticipation that miniaturisation via

the application of nanotechnology and new nanotools will lead to novel surgical and analytical tools and diagnostics for use ex vivo. It is envisaged that such techniques will increase our ability to identify predisposition to disease, monitor disease progression and identify the most relevant patient treatments. Moreover, a new generation of surgical tools is predicted that will be able to assist diagnosis and deliver therapy at a cellular level. In the context of larger ex vivo diagnostic devices, the focus of this research lies on nanointeractions.

DNA and protein chips are already widely used as research tools and are helping to provide a better understanding of the molecular basis of diseases and identify new molecular targets for therapy. It is perceived that many nanotechnology-derived tools will, in the future, be routinely used in diagnosis of many diseases long before they will be approved as treatments. In the case of these devices, nanoscale objects were defined as molecules or devices within a size range of one to hundreds of nanometres that are the active component or object, even within the framework of a larger micro-size device or at a macrointerface.

In the area of nanoanalytical techniques and diagnostic tools that will find application in the sectors of diagnostics, medical devices and pharmaceutical drug discovery, a wide range of technologies are being developed for both in vitro (diagnostics and sensors) and in vivo use with a range of biological targets including cells, DNA and proteins. Research and development in this field is extremely multidisciplinary and there is considerable synergy with the 'nanoimaging and manipulation' and the 'nanomaterials and nanodevices'.

Analytical and point-of-care diagnostic product design is already supported by the use of nanoparticles and nanodevices. For example, biosensor technology based on nanotechnology represents a huge opportunity to revolutionise diagnostics in the healthcare environment. Healthcare professionals in primary care and hospital clinics are shortly expecting to be able to use low-cost tests able to aid the diagnosis by simultaneous measurement of multiple parameters using a simple test strip without having to measure each parameter individually. Such test strips must be disposable and cost effective. So far only single analyte strips have been available but multisensor dry enzyme, hand-held systems being developed in Europe are leading the way towards fast and accurate multiparameter analysis.

There is an opportunity to focus on technically more mature developments that are more likely to be successful, and also to address niche markets that have better prospects for exploitation. Moreover, techniques and tools developed for analysis and diagnostics in the medical field have the potential for wider application, for example, in the context of environmental monitoring, control of food hygiene etc. While there is no lack of innovation in Europe, there was agreement that in many areas Europe is lagging behind. The DNA array market, for example, is dominated by Affimetrix, an American company. While DNA chips are being widely used in research, their difficulties in securing quality assurance and regulatory approval, their production flexibility and speed to result are well known. Although other more technically complex chips are available, so far they are substantially more expensive.

Underlying research was seen as a further European strength, particularly in relation to the identification of medical targets. Current genomics and proteomics techniques give a limited insight into cellular function. In the future, the combination of these techniques with imaging methods such as TOF-SIMS analysis, and raster electron microscopy by atomic force microscopy of living cells will give more information on individual protein function and cell signalling.

Imaging is becoming an ever more important tool in the diagnosis of human diseases. Both the development of imaging agents based on micro- or nanoparticles, organic dyes etc., and the development of highly sophisticated instruments supported by powerful computation have already given rise to a significant move from invasive to less invasive clinical imaging. This is allowing an ever more sophisticated imaging-based diagnosis, particularly in cancer, neurological and heart diseases.

Nanotechnology has an important role to play in further developing this field. Earlier diagnosis with a non-invasive tool can allow earlier treatment and this treatment approach can be much less expensive and often more effective. In addition, economists consider living healthier, longer lives to be economically beneficial. There are potential risks and side-effects from new imaging agents based on nanomaterials, but risk-benefit analyses will be/ are being conducted at an early stage.

One of the greatest problems in the transfer of these approaches into routine clinical use lies in the slow approval process of new materials for human use by regulatory agencies. Imaging at cellular, and even sub-cellular and molecular level, is still largely a domain of basic research. However, it is anticipated that these techniques will find their way into routine clinical use. Atomic force microscopy (AFM) and AFM-related techniques have become sophisticated tools, not only to image surfaces of molecules or sub-cellular compartments, but also to measure molecular forces between molecules. This is substantially increasing our knowledge of molecular interactions. Because of its interdisciplinary nature, the field of Nanoimaging and Manipulation is also identified as having a considerable overlap with the other disciplines, particularly in relation to nanomaterials and clinical, regulatory and toxicological issues. The field termed here 'Nanoimaging' overlaps with the field already called 'Molecular Imaging'. In this case the target to be imaged will have a spatial resolution in the order of 1-100 nanometres, and preferably a time resolution for imaging in the order of milliseconds. The opportunities for improvements and breakthroughs with the assistance of nanotechnology were seen in terms of both existing and emerging technologies.

Europe has a number of leading academic groups and several leading companies who are pioneers in the development of imaging techniques and innovative contrast agent production. Basic research has already developed the first methods to monitor in vitro the assembly of multicomponent biological complexes, protein trafficking and the interactions between single molecules. There is a recognised opportunity to use nanotechnology to improve these molecular imaging techniques, and to construct real-time intracellular tomography. Objective methods for assessment of image quality are also needed in vitro and in vivo. Tools are currently under development that allow in vitro evaluation of basic mechanisms, but it was felt that these could quickly be developed towards real ex vivo and clinical applications.

In the context of in vivo and clinical imaging, the development of novel techniques for macromolecular imaging was seen as a particular priority. Europe has been at the forefront of the development of polymeric gamma camera imaging agents and dendrimer-based MRI imaging agents. Improved image analysis is a particular goal. There is a need to improve 3D reconstruction and quantitative data analysis. Improved visualisation techniques are needed also for stereo-imaging, virtual and augmented reality imaging, and image-guided manipulation.

Opportunities exist for both invasive and noninvasive clinical imaging, e.g. endoscopy for targeted imaging, use of optical catheters and development of nanosized systems allowing manipulation, target selection, local stimulation and potentially local modification. Development of more sophisticated imaging equipment requires an integrated approach. Underpinning research must involve all aspects of the process. Parallel to the development of the analytical equipment, research and clinical development is ongoing to provide a new generation of nanoimaging agents. These include both synthetic nanoparticles (including dendrimers and polymeric nanoparticles) and biological nanoparticles (nano-organisms).

In the future it is likely that these imaging tools will become ever more complex, multicomponent systems combining contrast agents and tracking probes (e.g. quantum dots, magnetic and super-paramagnetic beads, nano-shells and nano-colloids) with new targeting ligands. For targeted systems, carriers can be used which may require additional surface modification, and linkers that bring additional challenges for physico-chemical characterisation and safety evaluation. In some cases combination of a range of signal modalities (e.g. organic dyes) is also used. Imaging and contrast agent design has a considerable overlap with nanomaterials and drug delivery system development.

6

Electronic Applications

The most pressing need for miniaturization on the nanoscale is in the field of consumer electronics. Current top-down lithographic techniques are reaching their fundamental limits. At some point, we will no longer be able to produce faster, more efficient, and denser microchips at an affordable price. Luckily, nanotechnology has a lot of promise in electronic applications. At the most basic level, nanoelectronics should theoretically use less power, generate less heat, and work more efficiently than its larger counterparts in the micro realm. These properties are predicted through known scaling laws, and have also been proven in a number of experiments.

The major problem right now is finding a way to make nanoelectronics feasible for consumer electronics. Our modern equipment happens to be fairly robust: cell phones can take a lot of damage before they break, and computers can work for several years before problems arise. Many proof of concept demonstrations in nanotechnology are not only extremely fragile, but they don't always respond well to normal atmospheric conditions.

Right now there are already a lot of different products on the market. For the most part, they are not really applications of nanotechnology. Instead, they are really just micro-electro-mechanical systems (MEMS). The interesting thing about MEMS is that much of the thinking that goes into it can later be used for some applications for the nanoscale version of the field. A number of exciting nanoelectronics demonstrations have already been published. Lasers, molecular switches, p-n junctions, logic gates, and early forms of memory have all been made in the lab. It will be some time before scientists and engineers can bring all these different components together into an integrated device. That is the true end-goal of nano-electronics research: a nano-computer.

Computing

The holy grail of nanoelectronics is an integrated computing device. In theory, such a computer would mimic the computational methods of present-day microchip processors. However, a nano-computer would take advantage of higher component density, low power requirements, and greater overall efficiency.

One obstacle stands in our way: we can't fabricate components with accuracy or precision. These two fundamental problems are the result of our poor control over the sizes and diameters of nanostructures and our inability to move them into position with ease.

On the accuracy side, we have the problem of large size distributions. With any given synthesis method, there is usually a non-uniform distribution of sizes that lead to differences in characteristics from structure to structure. In order to move further with nanoelectronics, we'll need tighter distributions. One way to guarantee this has been the traditional top-down approach, but lithography and other destructive techniques never provide the product density we need and also damages the final product. A top down technique introduces problems with placement. For integrated devices, each component must be in the right relation to its adjacent/connected components. We can't have components overlapping each other since each component must have proper isolation. Again, a top down approach isn't the best to achieve this.

It would appear then, that some bottom up method must be devised. On one side, we have our traditional semiconductor nanowire approach, and on the other, the molecular electronics approach. Both hold some promise in achieve the goal of a nano-computer.

For nanowires, it's obvious to researchers that it is the physical crossings (p-n junctions) that provide all the functionality. Their research has been focused on creating large numbers of crossings in a reproducible process. One such approach utilizes a Langmuir-Blodgett thin-film technique to apply layers of axially aligned nanowires on top of each other. Using this technique, it's possible to layer orthogonal sheets of nanowires over each other, thus creating thousands of crossings in a very small area.

Molecular electronics is focused right now on creating useful components. While nanowires have already been proven as building blocks for junctions, transistors, and logic gates, molecular devices have yet to demonstrate good numbers. Most of the problem lies in the donor-acceptor model that molecules must operate under. They are not semiconductors and hence do not conduct electricity in the normal manner. It will be some time before molecular electronics take off, but when they do it's quite possible that they'll be easier to manufacture do to the bio-mimicry that the science utilizes.

Electrical Components

Many important electronics applications require reliable

semiconductor materials. Using molecules is a novel way to tackle the problem, but it also introduces new problems that you won't run into when dealing with run-of-the-mill semiconductors.

Granted, individual molecules are smaller, but in terms of engineering they represent a nightmare. We haven't solved the problem of protein folding, and it will be some time before molecular electronics really gets rolling. In the meantime, it's up to semiconductor nanoelectronics researchers to come up with ways to do the same task with traditional nanotechnology structures like quantum dots and nanowires.

Most problems with these two structures are the result of our inability to quickly position or self-assemble them on a substrate. Quantum dots typically exhibit better self-assembly characteristics, but even then it's not on the level of what we'd need for an integrated device.

Let us discuss the developments in nanoelectronics so far. The fundamental component of electronics is the p-n junction, or diode. This is the piece in a circuit that restricts current flow in one direction only. The 'p' and 'n' refer to the semiconductor type (hole and electron, respectively). You need both kinds to make integrated devices. One way to make a p-n junction is to make a physical crossing between a p-type nanowire and an n-type nanowire. Another way to do it is to dope a single nanowire with p- and n-type regions. Either way, these basic components have already been fabricated and used as simple LEDs.

The next step in electronic device fabrication is the holy grail of digital logic: the logic gate. Digital logic is the governing principle behind all computations in electronics. On a single microchip, there are millions of logic gates. Researchers have successfully made nanowire logic gates by connecting several nanowire p-n junctions in series.

Nanowires and quantum dots have found many uses in electronics applications that are more traditional in comparison to the molecular electronics field. Though neither field has made significant breakthroughs yet, one or the other will likely find a key to unlock its true possibilities within the next few years.

LASERS

Contrary to popular belief, lasers are not used primarily as torture devices for secret agents who infiltrate evil complexes. Lasers are one of the most versatile tools in the realm of electronics. They are essential for a number of photonics applications like fiber optics. They're also useful as an analysis technique for a wide variety of materials.

The problem with lasers is that they're not that efficient. Yes, there are some really bright lasers out there, but they happen to draw a lot of power and generate quite a bit of heat. To complicate matters further, most of our lasers are based on some compound III-V semiconductor mixes like GaAs or InP. While these are great materials for creating a broad spectrum of frequencies, they are not compatible with a silicon process due to lattice mismatch.

All lasers require quantum confinement in at least one direction. The modern laser is based on a quantum well that confines electrons into a 2-D plane. While quantum wells are relatively easy to manufacture, they are not very efficient in terms of quantum efficiency. Hence, the input-output ratio isn't that high.

New structures in nanotechnology have opened the door for brighter and more efficient lasers. To start things off, a nanowire offers confinement in two directions. Large nanowire arrays have already been used to emit coherent light at ultraviolet (and many other) frequencies. The problem with nanowire lasers is that often they require another laser of a different frequency to 'pump' it with electrons. From a device standpoint this is not desirable, as we want lasers to work independent of other lasers.

The most exciting development of late has been the quantum dot laser. The q-dot offers the maximum confinement for the best efficiency. A number of innovative approaches have also provided new power sources for electron pumping. The best example of this is a quantum dot laser that is a matrix of quantum dots and an electrically active polymer matrix. Depending on the laser type, the polymer can either remove or donate electrons from the quantum dots. This is a critical step for laser operation.

Since lasers are now used in daily life for a variety of optical media, their fabrication in the nanometer scale will ultimately pave the way for a new class of devices and storage media. A large portion of work has already been completed, but an affordable nanoscale laser is still in development.

MEMS and NEOMS

Though Micro-Electro-Mechanical Systems (MEMS) are not actually part of the nanotechnology revolution, they will ultimately play a large role in its development. MEMS takes the idea of the lithographic semiconductor process and puts an entirely new spin on it. Lithography is used primarily as a way to pattern microchips with extremely fine detail. In general: the finer the detail, the more complex the computing system, the faster the microchip. But MEMS isn't about component density. The goal behind

MEMS is to create functioning electronic AND mechanical systems in the micrometer regime.

Making mechanical devices with micrometer features is not as simple as patterning a microchip with a 90 nm EUV litho progress. It requires great precision and innovation to make a micro-array of fully addressable mirrors, for instance. Big names like Texas Instruments have recently applied MEMS toward the field of optics. They've created an array of fully tilting micro-mirrors that are now used in modern projectors.

Such a micro-mirror array requires a high level of integration between extremely small mechanical parts and their electronic controls. Much of the work going into MEMS will pave the way for nanoscale applications, namely the field of nano-electro-optical-mechanical systems (NEOMS).

Consider the current micro-sensor. It is a MEMS application that sports a cantilever that is only micrometers long. This cantilever vibrates at a natural frequency dependent on its material and length. The tip of the cantilever can be fictionalized to bond with a target molecule. When that molecule lands on the tip and sticks, it changes the vibration frequency of the cantilever. This change is detectable, and the entire device forms the modern biological sensor. It's already been used to find a number of airborne agents like anthrax.

The micro-sensor idea serves as a good example because it could easily be transposed to a nanoscale version with even greater accuracy. With smaller cantilever sizes, even the smallest molecules could be detected with ease. Furthermore, all of the foundry work related to MEMS applications also finds use in NEOMS applications. Though right now, MEMS is more advanced than the state of NEOMS, one day we'll begin to see a wide array of funky new devices that take advantage of tiny mechanical parts that are only nanometers in length.

One possible realization of nanoscale electronics has long been a mainstay in science fiction. It's known as molecular electronics, and it utilizes a host of existing molecules to do the work. The reason why molecular electronics is so attractive is mainly because a majority of the legwork is already done: a large number of known donor/acceptor molecules have already been synthesized over the last 50 years. Furthermore, the invention of the STM and AFM in the 1980's have given researchers unprecedented control over the placement of molecules.

Classical thinking dictated that molecular electronics was an impossible endeavor. Using the basic equation of resistance of a wire, it would seem that a molecular wire would have near-infinite resistance (hence it could not conduct electricity). But early theorists did not foresee the development of charge transfer molecules developed in the mid-20th century. These donor-acceptor systems proved vital in creating a new class of electrically active solid-state materials. As individual molecules, however, researchers knew that such systems could be used one day for a molecular device. They just didn't have the right microscopes yet.

The reason why molecular electronics is so important is because it represents one realization of nanoelectronics using an existing toolbox. Chemists already know how to tailor donor or acceptor molecules. They are easy to synthesize and can self-assemble through a variety of functionalization techniques. Molecules are ideal because they're already in the nanoscale, with larger molecules approaching 100 nm. In terms of mass production, molecular electronics may be the only way to effective create large quantities of nanoelectronics.

Several research groups have already created functional molecular switches and molecular transistors. There are plans in the works for new kinds of memory that could store the entire contents of the Internet on a single disk. Eventually, molecular electronics will create cheap, ultra-fast, and efficient microchips that will power the next generation of computers.

Applications of Nanoelectrnic Devices

Complementary metal–oxide–semiconductor (CMOS), is a major class of integrated circuits. CMOS technology is used in chips such as microprocessors, microcontrollers, static RAM, and other digital logic circuits. CMOS technology is also used for a wide variety of analog circuits such as image sensors, data converters, and highly integrated transceivers for many types of communication.

CMOS is also sometimes explained as complementary-symmetry metal–oxide–semiconductor. The words "complementary-symmetry" refer to the fact that the typical digital design style with CMOS uses complementary and symmetrical pairs of p-type and n-type MOSFETs for logic functions.

Two important characteristics of CMOS devices are high noise immunity and low static power supply drain. Significant power is only drawn when its transistors are switching between on and off states; consequently, CMOS devices do not produce as much heat as other forms of logic such as TTL. CMOS also allows a high density of logic functions on a chip.

The triple compound "metal–oxide–semiconductor" is a reference to the nature of the physical structure of early (and interestingly now, the very latest) field-effect transistors, having a metal gate electrode placed on top of an oxide insulator, which in turn is on top of a semiconductor material. Instead of metal, current gate electrodes (including those up to the 65 nanometer technology node) are almost always made from a different material, polysilicon, but the terms MOS and CMOS nevertheless continue to be used for the modern descendants of the original process. Metal gates have made a comeback with the advent of high-k dielectric materials in the CMOS transistor as announced by IBM and Intel for the 45 nanometer node and beyond. The combination of MEMS sensors with digital signal processors on one single CMOS chip is sometimes known an CMOSens.

CMOS circuits were invented in 1963 by Frank Wanlass at Fairchild Semiconductor. The first CMOS integrated circuits were made by RCA in 1968 by a group led by Albert Medwin. Originally a low-power but slow alternative to TTL, CMOS found early adopters in the watch industry and in other fields where battery life was more important than speed.

Some twenty-five years later, CMOS has become the predominant technology in digital integrated circuits. This is essentially because area occupation, operating speed, energy efficiency and manufacturing costs have benefited and continue to benefit from the geometric downsizing that comes with every new generation of semiconductor manufacturing processes. In addition, the simplicity and comparatively low power dissipation of CMOS circuits have allowed for integration densities not possible on the basis of bipolar junction transistors.

Standard discrete CMOS logic functions were originally available only in the 4000 series (RCA "COS/MOS") integrated circuits. Later many functions in the 7400 series began to be fabricated in CMOS, NMOS, BiCMOS or another variant.

Early CMOS circuits were very susceptible to damage from electrostatic discharge (ESD). Subsequent generations were thus equipped with sophisticated protection circuitry that helps absorb electric charges with no damage to the fragile gate oxides and PN-junctions. Still, antistatic handling precautions for semiconductor devices continue to be followed to prevent excessive energies from building up. Manufacturers recommend using antistatic precautions when adding a memory module to a computer, for instance.

On the other hand, early generations such as the 4000 series that used aluminum as a gate material were extremely tolerant of supply voltage variations and operated anywhere from 3 to 18 volts DC. For many years, CMOS logic was designed to operate from the industry-standard of 5 V imposed by TTL.

By 1990, lower power dissipation was usually more important than easy interfacing to TTL, and CMOS voltage supplies began to drop along with the geometric dimensions of the transistors. Lower voltage supplies not only saved power, but allowed thinner, higher performance gate insulators to be used. Some modern CMOS circuits operate from voltages below one volt.

In the early fabrication processes, the gate electrode was made of aluminum. Later CMOS processes switched to polycrystalline silicon ("polysilicon"), which can better tolerate the high temperatures used to anneal the silicon after ion implantation. This means that the gate can be put on early in the process and then used directly as an implant mask producing a self aligned gate (gates that are not self aligned require overlap which increases device size and stray capacitance). Considerable research that has gone into using metal gates has led to the announcement of their use in conjunction with the replacement the silicon dioxide gate dielectric with a high-k dielectric material to combat increasing leakage currents.

CMOS (complementary metal–oxide–semiconductor) refers to both a particular style of digital circuitry design, and the family of processes used to implement that circuitry on integrated circuits (chips). CMOS logic on a CMOS process dissipates less energy and is more dense than other implementations of the same functionality. As this advantage has grown and become more important, CMOS processes and variants have come to dominate, so that the vast majority of modern integrated circuit manufacturing by dollar volume is on CMOS processes.

Structure of CMOS

CMOS logic uses a combination of p-type and n-type metal–oxide–semiconductor field-effect transistors (MOSFETs) to implement logic gates and other digital circuits found in computers, telecommunications and signal processing equipment. Although CMOS logic can be implemented with discrete devices (for instance, in an introductory circuits class), typical commercial CMOS products are integrated circuits composed of millions (or hundreds of millions) of transistors of both types on a rectangular piece of silicon of between 0.1 and 4 square centimeters. These bits of silicon are commonly called chips, although within the industry they are also referred to as die (singular) or dice (plural).

In CMOS logic gates a collection of n-type MOSFETs is arranged in a pull-down network between the output and the lower-voltage power supply rail (often named Vss or quite often ground). Instead of the load resistor of NMOS logic gates, CMOS logic gates have a collection of p-type MOSFETs in a pull-up network between the output and the higher-voltage rail (often named Vdd). Now pull-up and pull-down refer to the idea that the output node, which happens to be where the pull-up and pull-down networks intersect, exhibit some internal capacitance that is charged or discharged respectively through pathways formed by the p/nMOS networks for various inputs.

This capacitance is charged when there is a direct path from Vdd to the output, and discharged when there is a direct path from output to ground. Notice that a digital CMOS circuit cannot (ideally) be in a pull-up and pull-down phase at the same time, or else both the p/n-networks will fight to keep the voltage on the capacitance either Vdd or ground. The p-type transistor network is complementary to the n-type transistor network, so that when the n-type is off, the p-type is on, and vice-versa.

CMOS logic dissipates less power than NMOS logic because CMOS dissipates power only when switching (dynamic power). On a typical ASIC in a modern 90 nanometer process, switching the output might take 120 picoseconds, and happen once every ten nanoseconds. NMOS logic dissipates power whenever the output is low (static power), because there is a current path from Vdd to Vss through the load resistor and the n-type network.

P-type MOSFETs are complementary to n-type because they turn on when their gate voltage goes sufficiently below their source voltage, and because they can pull the drain all the way to Vdd. Thus, if both a p-type and n-type transistor have their gates connected to the same input, the p-type MOSFET will be on when the n-type MOSFET is off, and vice-versa.

Single Electron Tunnelling (SET) Devices

A SET device is a three terminal device based on the Coulomb blockade where the number of electrons on an island or dot is an integer number controlled by a gate. The dot may have up to thousands of electrons depending on the size and material. It may be produced in a number of different ways, the most common are metallic islands or semiconducting quantum dots. The basic operation requires an island of electrons with a capacitance C which is small enough that a charging energy for the island, (e^2/C) is much larger than the thermal fluctuations in the system, (k_BT).

Electrons may only flow through the circuit by tunnelling onto the first unoccupied energy level, μ_{N+1}. Therefore, electrons will only flow one by one if the bias voltage, V is increased such that $\mu_l > \mu_{N+1} > \mu_r$ or a gate is used to change the electrostatics of the island to produce the same tunnelling conditions. Numerous SET based memory circuits have been suggested and a number demonstrated at helium temperature and at 40 K, which are aimed at, low power, rather than high-speed applications. The first uses binary decision diagrams to produce an AND gate while the second uses oscillatory characteristics of a multiple gate SET transistor to produce a XOR gate.

SET devices are at present believed to be useful predominantly for memory, electrometer and metrology applications. To make a SET device operational at room temperature, it is estimated that the charging energy of the island, (e^2/C) should exceed the thermal energy k_BT by at least a factor of 10. This suggests that the island of the SET device must be of the order of 10 nm. For reliable circuit operation, however, e^2/C should exceed the thermal energy k_BT by a much larger factor and hence the feature sizes must be smaller than 10 nm.

Simulations of complete SET circuits using a conventional type of architecture and incorporating perturbation by background charge fluctuations suggest that it will be necessary to go to dimensions of the order of 2 nm, and that liquid nitrogen cooling may be necessary. The feature dimension will therefore depend on the control of the background charge fluctuations. Recent calculations suggest that quantum dot array structures (such as multiple tunnel junctions) are less susceptible to disorder and background charge effects.

The energy needed to read or write a bit and the frequency of the circuits are limited by the uncertainty principle, $\Delta E \ \Delta t \geq h$. To prevent bit errors, the circuit cannot operate too close to the minimum uncertainty product. The quantum limit is then approximately given by $E/f = 100\ h$. Here E is the energy needed to write a bit and f is the clock frequency. SET based circuits operate at the quantum limit. The trend in SET devices is to increase E and f simultaneously so that as the speed and operating temperature increase, the ratio E/f remains constant and the circuits stay at the quantum limit (figure 1). CMOS circuits operate far above the quantum limit but are approaching it as E decreases and f increases (figure 1).

There are three important limits, which determine the ultimate performance of such systems, in particular the thermal limit, the quantum limit, and the power dissipation limit (figure 1). The energy necessary to write a bit determines the thermal limit. This energy must be bigger than the average energy of the thermal fluctuations, k_BT, otherwise bit errors will occur. For the present SET circuits, this energy is 10^{-22} J (10^{-3} eV) which corresponds to a temperature of 10 K. The trend in SET circuits is to increase this energy and thereby to increase the operating temperature of the circuits.

The optimum value for the energy to write a bit for room temperature operation is about 4×10^{-19} J (2 eV), which is a factor 100 greater than k_BT. In principle, the speed of SETs is limited by the RC time constant that, for

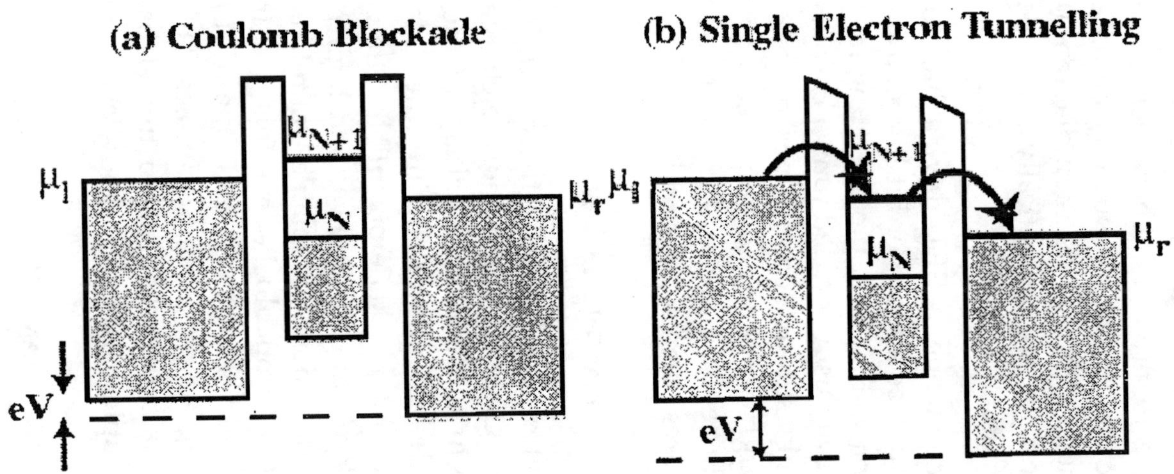

Figure 1. For an island of total capacitance C with N electrons, being μ_N the chemical potential of the highest filled electron state, μ_{N+1} the chemical potential of the first available empty state for an electron and μ_l and μ_r the chemical potentials of the left and right electrodes respectively, it may be shown that the energy to add an electron to the island is $\mu_{N+1}-\mu = e^2/C$.

capacitances of 1 aF, corresponds to a switching speed of 0.1 ps. To take advantage of these speeds, however, the logic architecture would have to be local so that the SETs would not have to drive a high capacitance line across the chip. Logic circuits would possibly be based on local architectures, such as binary decision diagram (BDD) logic or cellular automata, are theoretically and experimentally under investigation.

In practice, when a SET device has to drive an external load, such as a word or bit line in a memory cell, there are RC delays that limit the operating frequency and it is likely that even using graded tunnel barriers in a Yano type device, only sub-ns access times may be achieved. Due to the high impedance required for Coulomb blockade, SET devices are easier to implement into memory structures, than logic circuits.

Nano-flash Devices

Nano-flash devices are basically three terminal devices where a floating gate is charged and the charge produces a large change in the threshold voltage of the transistor channel. The design allows an intermediate between DRAM and Coulomb blockade potentially allowing higher density than DRAM at lower power and higher operating temperatures. In addition, nonvolatile Dram-like memories based on the Coulomb blockade effect are intensively investigated. Both Hitachi's PLED and Likharev's NOVORAM being are prominent examples. The key issue is the creation of extremely flexible tunnel barriers, for instance by multiple barriers or sandwiched barriers.

Yano Memory

The Yano type memory is a 2 terminal device where information is stored in deep traps in poly-Si. The devices are created on a 3 nm thick Si film using 0.25 μm technology where one or more dots are formed naturally in the vicinity of a FET in which trapped charge modulates the threshold voltage of the FET. The device can be operated at room temperature and has been integrated in very large-scale memories although it is not certain if Coulomb blockade is of any relevance for device operation. One of the major problems of this type of memory is relying on the natural formation of dots and the resulting poor control of device characteristics. This may be a major hurdle to manufacturability.

The advantage is a small cell size of $2F^2$, one quarter of a folded-data line DRAM cell size. Since 1994 at least three major companies have introduced technologies for room temperature Coulomb blockade memory cells. They are compatible with CMOS process and integration on 250 nm technology level was demonstrated in one case. Their properties place these memories between today's DRAM and flash EEPROM. Hitachi presented the first single electron based integrated circuit by making an 8 x 8 memory cell with read / write operation. The operation voltage is 15 V and the device is based on ultra thin polysilicon wires (3 nm x 100 nm) in which the memory node consists of an isolated poly grain representing a potential well. The presence of charge in this well modulates the conductance of naturally formed current paths between the grains. Because of the compatibility with "classical" silicon processing these results offer a real breakthrough. The device operating principle, however, relies upon the statistics within the poly wire.

Resonant Tunnelling Diodes (RTDs)

Resonant tunnelling is a device concept whose inherent multistability allows for very compact circuit design (not hindered by a need to go much below a 200 nm design rule), and its inherent speed paves the way into the GHz regime. The basic concept of the RTD is illustrated in Figure 2. Negative differential resistance (NDR) is produced in a double barrier structure with a resonance peak at some voltage, V which corresponds to resonant tunnelling of electrons through a subband energy, Eo in the quantum well between the barriers.

Multiple NDR devices may also be produced by designing systems with multiple subbands or extra wells. The RTD is normally a vertical tunnelling transistor with the vertical dimensions produced by growth (i.e. MBE or CVD) and the lateral dimensions produced by lithography, although some lateral tunnelling transistors have also been demonstrated. The dimensions in the tunnelling direction are therefore typically a few atomic layers thick while the lateral dimensions are restricted by lithographic sizes. The width of the devices determines the current and hence the power dissipation. Therefore smaller widths typically produce lower power devices.

RTD device technology is now quite mature with many demonstrations of circuits. These are the first quantum transport devices to have made it into (pilot) production. The design of the RTD depends on the specific application in question. To reduce the power in memory and logic, one of the important issues is to produce the minimum valley current density possible at a low voltage. The peak to valley ratio (PVR) must be large enough to allow an appropriate memory or logic function with a reasonable noise margin; but a too large PVR would result in a slower circuit as it takes too long to reach the peak current density.

The switching time of the RTD depends on the RC time constant of the device and the charging time of the RTD is determined by its peak current density. Hence there

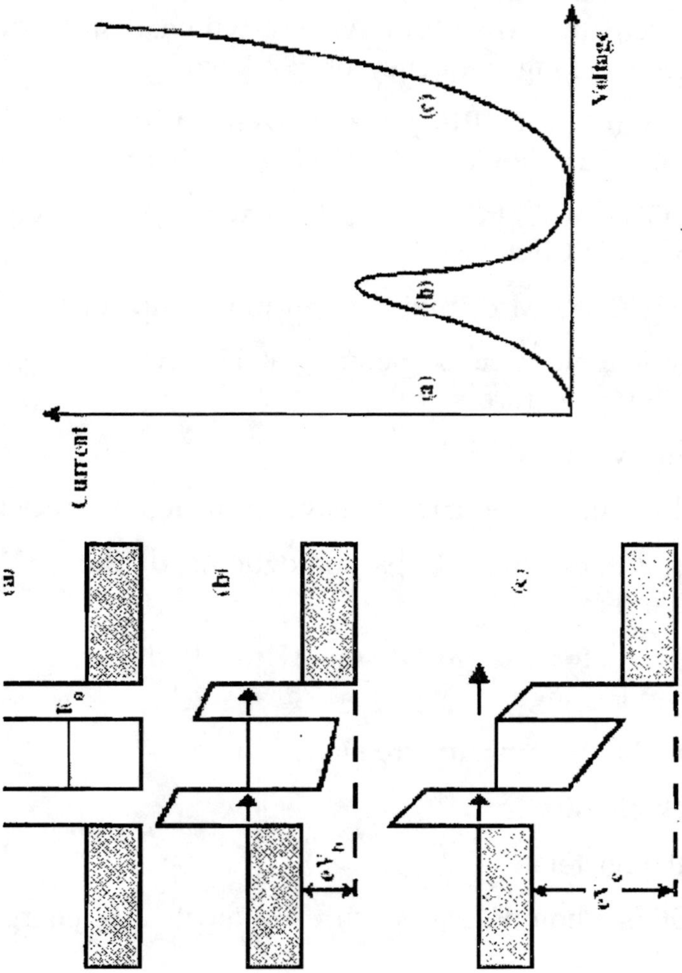

Figure 2. Basic concept of the Resonant Tunnelling Diodes.

is a trade-off between high speed and low power in the design of the circuits. For high speed logic applications, higher peak current densities are required, typically >10 kA/cm^2 allowing a maximum clocking frequency of 6.25 GHz compared to the low power TSRAM where the peak current densities are <0.16 A/cm2 which results in a maximum clocking frequency of 592 kHz.

Resonant tunnelling has proven fruitful in III-V devices and has shown the following results: for Devices:

— 712 GHz InAs/InGaAs oscillator with 0.3 μW output power for Circuits:

— 50 nW TSRAM cell with a 150 μm^2 footprint

— generic logic circuits operating at 12 GHz with 20 μm minimum feature size

— memory circuits

— multivalued logic circuits have been demonstrated

— monolithic 4-bit 2 Gsps analogue to digital (ADC) converters

— 3 GHz clocked quantisers with 40 dB spur free dynamic range

— 40 GHz static binary frequency dividers

— clock circuits

— shift registers

— 2 GB/s photodetectors with low switching energies of 30 fJ.

Rapid Single Flux Quantum Logic

Rapid Single Flux Quantum (RSFQ) Logic is based upon a superconducting quantum effect, where a flux quantum is used as a bit. The basic switching elements are

Josephson junctions (JJ). Being a quantum device, RSFQ does not suffer from the 200 nm optical lithography barrier. RSFQ is the only known technology in which circuit speeds above 100 GHz and power dissipation down to 1 μW/gate can be envisaged. There are two basically different technologies to produce RSFQ circuits, depending on wether low temperature superconductors (LTS) or high temperature superconductors (HTS) are employed. Due to its superconducting principle, a RSFQ device needs cooling and it should be remembered that the operation temperature of the device is lower than the critical temperature of the bulk superconductor material. The availability of adequate cooling systems, which comply with needed specifications in the limits of reasonable prices, is one of the most important drawbacks for the market introduction of this technology.

Low Temperature Superconductors

The LTS technology uses superconductor-insulator-superconductor (SIS) tunnel junctions. The maturest technology is a three-layer process using Nb as supercorductor and AlOx as insulator. Nb-based devices must be cooled down to liquid helium temperature. Circuits with several thousand junctions, produced with a 3.5 μm design rule, and frequencies up to 40 GHz can be obtained commercially. As the critical dimensions are limited by the magnetic penetration depth, Nb miniaturisation should be possible down to 100 nm. An alternative is to use NbN as superconductor, where circuits can operate at 10 K.

Current circuits have up to 1000 junctions, i.e. one order of magnitude less than the conventional Nb technology. This technology is less mature mainly due to the lack of self limiting tunnel barriers. For most current

LTS circuits 10 mW (@ 4K) of cooling power is sufficient. In terms of input power this corresponds to about 1 kW of electrical power, which makes LTS suited only to large systems where the energy consumption and size of cooling engines is of minor importance or to applications exploring quantum effects of superconductors, such as in metrological standards. Very large computers such as the Teraflop / Petaflop (10^{12} and 10^{15} floating point operations per second, respectively) computer, are examples of the first case. They would consume enormous amounts of power if realised in silicon, while the use of LTS reduces the cooling problem and offers a more compact layout.

High Temperature Superconductors

HTS technology uses mainly YBCO (or other 123 cuprates) as superconductor and mostly PBCO as insulator. Presently, only simple circuits, such as shift registers, with less than 30 junctions have been demonstrated to operate in the order of 40 GHz and at temperatures below 50 K. This is two orders of magnitude smaller complexity than current LTS technology. Due to the fact that YBCO is an anisotropic material, the fabrication is far more complex than with LTS. There are diverse techniques to produce HTS Josephson junctions, such as using the interface of bicrystals, caxis microbridge and ramp type junctions. Which fabrication method will be chosen is not clear as all of them have pros and cons.

For example, ramp type junctions have the best reproducibility, while microbridge systems are more easily scalable, a factor that may become important in the future. The advantage of HTS in respect to LTS is that it is supposed to operate at higher frequencies (up to 500 GHz) and higher temperatures (50K), when the circuits are produced with a 500 nm groundrule. The feature size 500 nm is limited by the magnetic penetration depth. The higher temperature allows for less extensive cooling efforts, while the higher frequency may open a window for new applications. As HTS circuits will be operated at higher temperatures than LTS it is expected that the risk of bit error rates increases. Measurements of these values have not been performed.

Intramolecular Nanoelectronics

The term intermolecular nanoelectronics is used to distinguish switching at the single molecule level compared to the molecular electronics at the bulk, large molecular number level as used in displays. Aviram and Ratner proposed in 1974 that suitable molecules could be used as functional electronic components, specifically for rectification. Ashwell demonstrated such rectification behaviour using zwitterionic molecules and a magnesium evaporated electrode. While only a few major results have appeared in the field, the potential for mass selfassembly of circuits using molecular sized switching elements by cheap chemical or biological reactions is one of the few technologies which has the potential to match the ultimate densities of CMOS.

In particular, intramolecular nanoelectronics is attempting to reduce logic and memory elements of circuits to the sizes of molecules and seeking:

— ultimate bit densities of 10^{12} bits/cm^2

— decreased switching cycle times to 10 ps o reduced energy per bit cycle to 1 eV

Intramolecular nanoelectronics is still in its infancy, but holds promise in that it potentially provides a bottom-up way to produce sub-nm sized functional devices without the inherent tolerance problems of a lithographic (top-down) approach. First manufacturable solutions are predicted to be bolt on technologies on top of a CMOS chip with the CMOS chip providing input/output and communication to the rest of the world.

Molecular Electronic Switching Devices

Among molecular switches, electric-field controlled molecular electronic switching devices are closest to conventional semiconductor devices and therefore the most likely candidates for applications. The basic idea is to use the central part of molecules for functions while the endgroups are used to self-assemble. Diblock copolymers, for example, may be self-assembled into chains and purified to very high yields. Thiol endgroups can be used to anchor the copolymers to metal electrodes.

The mechanism for the functional behaviour can be very diverse. As already indicated, attaching electrophile and electrophobic groups to either end of the molecule induces rectifying behaviour. Aviram proposed a number of molecules that switch between conducting and insulating states by oxidation/reduction. One example is tetrathiafulvalium (+) with a partially filled highest occupied molecular orbital to allow free states for conduction and tetrathisfulvalene which is an insulator because of its filled highest molecular orbital. The transport properties are very similar to Coulomb blockade.

Current voltage data from single designer molecules have been demonstrated such as the rectification properties of a benzene ring attached to two gold electrodes using S atoms (Benzene-1, 4-dithiolate). More complicated structures have also been demonstrated where the rotation of a single bond of the p orbitals allows the conductivity to be changed by over 6 orders of magnitude, providing a switching mechanism in a polymer chain. The first "molecular transistor" actions were shown by R.P. Andes using selfassembled gold particles on top of a,a'-xyldithiol (XYL) molecules on an Au(111) surface. The second electrode was a STM tip and a Coulomb staircase function was demonstrated at room temperature. A second demonstration was using carbon nanotubes placed across two electrodes with a controlling back-gate.

Clear Coulomb blockade and also single electron transistor properties were demonstrated at room temperature. In semiconducting nanotubes, "classical" field effect transistor action could be obtained. These experiments demonstrate the basic transistor principles but neither technique is suitable for mass manufacture. A number of proposals have appeared in the literature for molecular devices but few have been demonstrated.

Recognition of conformation and adaptation of individual molecules has been demonstrated where the STM is used to understand bond rotations in molecules and to determine how they adapt or can switch their state. Such conformational processes are used in natural systems and form a basis for molecular switches. It has been demonstrated that molecular systems such as fullerene based systems can be incorporated in two and three terminal devices and that interesting electronic characteristics can be obtained, i.e. macroscopic quantum state in a bucky tube.

The area of self-assembly has for some time been subject to designer molecular concepts through optimisation of molecular chain length and omega functionalisation as well as techniques such as contact printing to define molecular patterns on the scale of tens of nanometers. Lehn's supramolecular concepts have also been explored in the design and self-organisation of grid-like metal co-ordination arrays using molecular systems. The assembly of devices and arrays may be expected to be radically different from current approaches. For instance, at the macroscopic scale chemical interactions can be used for the three dimensional self-assembly of millimetre-scale components. A different form of molecular recognition is to use DNA to create a binding interaction between nanocrystals of colloidal metals in solution (or potentially at a surface).

Electromechanical Switching

Several illustrations of STM based molecular manipulation have in recent years been proposed as a means for performing electronic computation. The main consideration of electromechanical switching at the molecular level is that the properties are controlled by deforming or reorientating a molecule rather than moving the electrons on the molecule. An example is the C_{60} based single molecule electromechanical amplifier, which has been demonstrated by deforming the molecule using a STM tip.

The device relies on the controlled vertical deformation of the C_{60} cage resulting in mechanical modification of resonance tunnelling bands. This demonstrates a fundamentally new approach to switching and amplification using molecular mechanics and quantum processes. The speed of such a device would be limited by the vibrational frequency of C_{60} at over 10 THz (10^{13} Hz) although much lower frequencies (10 Hz) have been demonstrated. Another example is the Molecular Abacus device using the fullerene C_{60} as beads, a 0.25 nm high monatomic step as the rod and STM tip as the finger to both reposition in one dimension and count by imaging.

The abacus itself has an active area defined as around 1 nm x 13 nm and is a molecular mechanical device. By moving a single atom out or into an atom wire researchers at Hitachi have simulated a two state electronic device of atomic dimensions. The device is called an atom relay or molecular relay. Two-state devices based on the change of a molecular conformation can also make a relay such as a rotamer.

Photoactive/Photochromatic Switching

A number of molecules, specifically a number of proteins, may have their electron distribution changed by the absorption of photons to produce switching effects. The biological photochrome bacteriorhodopsin has been suggested as one possible type of molecule for holography, spatial light modulators, neural network optical computing, nonlinear optical devices, and optical memories. The significance of bacteriorhodopsin stems from its biological function as a photosynthetic proton pump in the bacterium Halobacterium halobium. A combination of serendipity and natural selection has yielded a native protein ideal for optoelectronic applications. Other examples of optoelectronic biological molecules include visual rhodsin, chloroplasts and photosynthetic reaction centres.

Molecular Wires

A further important problem, which has escaped much attention, is that most polymers are very poor conductors and hence molecules must be found with good conducting properties if circuits are to be built up. Note that utilising the delocalised electrons on aromatic and acetylene groups is not sufficient to obtain good conducting properties as required by the microelectronics industry in circuits. The delocalised π electronics in such compounds occupy the "homo" (highest occupied molecular orbital) fully, so that for charge transport an extra electron must be added to the "lumo" (lowest unoccupied molecular orbital), which is typically a few eV away. Thus, the compound acts as a semiconductor and good conducting properties can only be obtained by doping the systems. The only metallic wire like molecules that have been studied in more detail are carbon nanowires, where the bandstructure allows metallic behaviour. These compounds seem quite promising for some applications, but ways must be found to connect them, through conducting junctions at specified positions, to the active switching element. The RNA synthesis pursued by the Technion group could be a useful approach in this direction.

Spin Devices

Magnetic RAMs (MRAMs) are nonvolatile memory (NVM) devices whose operating principle rest upon the differences in conductance of multilayered stacks of nanoscale magnetic materials. Several concepts have been proposed in the past including giant magnetoresistance (GMR) devices (functioning on the basis of ferromagnetic/antiferromagnetic coupling) or device concepts based on the magnetic behaviour of the electron. The magnetic state does not require external power to be stabilised and hence applications are in NVM or static RAM. Since spin is already defined on a single electron, the devices should be able to downscale to very small dimensions. MRAM circuits are built on existing CMOS and use the GMR effect in magnetic multilayers. The effects exploited in MRAM are based on the differences in transport properties and density of states for spin-up and spin-down electrons in a magnetic metal. This difference in properties of minority and majority spins in a material with a high number of available carriers for extremely small structure sizes (<10 nm) triggers the quest for new device concepts. In principle, MRAMs are an array of ultra-small magnetic sensors with hysteresis constituting the bits.

The bits are "set" by a magnetic field (order 10 Oe) generated by 2 crossing conductors carrying a current in the order of 10 mA. The bits consist of magnetic thin multilayers exhibiting magnetoresistance, which enables one to read their memory state through measurement of their resistance (read-action). The signal magnitudes (order 5 mV) are read by Si sensing circuits using auto-zero schemes to level out the transistor imbalances and enable correct reading of the small signals.

MRAM based on magnetoresistive effects in metallic structures presents some advantages compared to existing NVM approaches. Scaling rules of MRAM are dictated by Si-technology. New materials development (e.g. tunnel magnetoresistance) will improve density, speed and power performance. Access times of the order of 300 ns have been reported so far, but it should be possible to reach10 ns (at room temperature) within a few years. VS Nonvolatile Electronics Inc. demonstrated MRAM with 200 nm cells producing 8% output in resistance change with a clear memory effect. For a future implementation of the memory circuit some additional requirements have to be met: careful control of the interlayer thickness, thin insulation, a magnetic keeper structure and conductor materials with higher allowed current densities ($>10^7$ A/cm^2). Analysis showes that sense line width less than 100 nm and array densities of 5×10^8 bits/cm^2 are possible with GMR. Honeywell Inc. has been pursuing the AMR-MRAM route since 1984 (based on the Anisotropic Magneto-Resistance effect in permalloy NiFe) and foresees the transition to GMR elements to result in 4 Gbit/cm^2 NVM technology. Their objective is full CMOS compatibility to ensure lowest power, highest reliability and rapid insertion into existing markets.

Lower cost can be achieved in MRAM compared to the equivalent semiconductor memory due to the need for only 2 to 3 metal interconnect levels. Also, because of the compact architecture process feature sizes that are 1 to 2 generations behind semiconductor memory / DRAM can be applied to achieve competitive densities. A new development in the MRAM area is the use of memory nodes based on Tunnel Magnetoresistance (TMR). In brief, these components are a tri-layer structure consisting of two ferromagnetic metals, spaced by a thin tunnel oxide. Similar to spin-valve structures, these tri-layers show a difference in resistance dependent on the alignment of the moments in the ferromagnetic layers.

The main difference between TMR structures and the presently used all-metal GMR structures is their much higher resistance and higher (relative) resistance change in the low-field regime. Further, TMR devices are compatible with CMOS. Most of the successful TMR devices have been fabricated using shadow masking technology. To enter the sub-micron regime and achieve IC-compatibility standard processing techniques can be applied, 100 x 100 nm^2 in industrial laboratories.

7

Industrial Applications

The present trend towards miniaturisation—the competitive quest for ever smaller machines and components that use less resources (e.g. energy, materials) —while offering the potential for the cheap mass-production of increasingly complex goods is rapidly pushing industry into the nanometre realms.

The coming era of nanotechnology is being made possible by the remarkable convergence of many technological advances in this decade that include:

— hugely powerful computers that allow the design of new materials and the simulation of their properties

— a new generation of microscopes that can provide images at the nanometre scale, as well as measure and manipulate atoms and molecules

— the advent of virtual reality that enables us to visit and experience the wonders of this new and hitherto unimaginable nanoworld

As the work of scientists tends increasingly towards the size domains of molecules and atoms, the very nature of matter is increasingly close to being controlled and manipulated. The possibilities this opens up are endless: from the production of new, lighter and stronger materials which have applications in areas as diverse as space travel to bone implants, to the creation of therapeutic drugs with individual-specific properties.

Nanotechnology can best be considered as a 'catch-all' description of activities at an almost vanishingly small scale that have applications in the real world. A nanometre is a billionth of a metre, that is, about 1/80,000 of the diameter of a human hair, or 10 times the diameter of a hydrogen atom.

An early promoter of the industrial applications of nanotechnology in the UK, Professor Albert Franks, defined it as 'that area of science and technology where dimensions and tolerances in the range of 0.1nm to 100 nm play a critical role'. This definition neatly allows for both the 'bottom up' and 'top down' approaches* to nanotechnology. It encompasses precision engineering as well as electronics, electromechanical systems (such as the development of 'lab-on-a-chip' devices) and mainstream biomedical research and development in areas as diverse as gene therapy, drug delivery and novel drug discovery techniques.

The role of nanotechnology as a major driver of technological change and its consequent importance in the shaping of world economies in the next millennium is undisputed, as evidenced by the commitment to nanoscale research by the US and Japan. It is crucial that the UK also identifies those technologies that may offer the most economic benefits. A recent survey has demonstrated that the market for microengineered products is greatest in electronics and biomedicine. These conclusions can in very general terms be extrapolated to nanotechnology. The UK is not an owner of electronics or semiconductor technology in the main, but tremendous opportunities exist in integrating these technologies into systems or components in high added value niche market products for biomedical and related applications. These applications include medical and environmental sensors, diagnostic and analytical devices, and range of equipment for minimally invasive surgery. There may be also be some potential for low cost, high volume applications in these areas. There are also important opportunities for the UK in the manufacture of the machine tools that industry will require in order to make products with nanoscale components or dimensions.

It is essential that a strategy for nanotechnology be developed as the basis of the revitalisation of economy. This is particularly urgent as our service sector is already also under threat. Advances in information technology are enabling competitors to move in to this sector, viewing it as another commodity industry - and one that requires fewer people.

The advent of the nanoindustrial revolution means that scientific and engineering disciplines can no longer justify their fondly preserved 'pigeonhole' existences - and this applies equally to University departments, research councils and learned scientific institutions, as well as companies. *Integration* is now the name of the game. Solutions to the industrial and social problems of the 21st century will be solved fastest and most effectively by multidisciplinary teams of nanoscientists that have been trained to understand each others 'language', and future business and education strategies need to take account of the need to develop these teams.

Manufacturing

Infrastructure Requirements

A wide range of production capabilities, training and

facilities are required as part of the creation of an infrastructure that will nurture nanotechnology and provide the basis for industrial development. For example, mathematics, computer modelling and simulation skills will be essential as well as an understanding of tools and standards. Frontier research requires advanced instrumentation to be available across the board; from the level of individual laboratories to national facilities. There is also a need for research on state-of-the-art instruments and their deployment

Key issues are:

— the production of, or access to specialist materials

— the adoption of advanced manufacturing processes

— access to specialist tools needed for manufacturing, test, assembly and inspection

— the installation of ultra-clean manufacturing facilities

— the provision of adequate training facilities for the development of skilled manpower

It is worth noting that business opportunities will exist at all stages of development of the new technology, including the provision of the basic requirements.

Materials for the Nanoindustrial Revolution

Materials science and technology is fundamental to the majority of the applications of nanotechnology. 'Raw' materials such as semiconductors, oxides and specialist organic and inorganic chemicals, will need to meet new specifications and parameters. For example.

— *Nanoparticles*: Controlled production of particles in the 1 - 100 nm size range is crucial, and handling of these fine particles will be a key issue.

— *Quantum structures*: Material purity is of the highest importance here, and research into production methodology is required.

— *Multilayer thin films*: These require clean deposition equipment and environment (impurities and defects will ruin the properties of the films) with fast turn-around and high throughput. Also, very high purity materials will be needed for sputtering and evaporation sources.

— *Nanomechanical devices*: The physical integrity of the material used to produce the devices will be of key importance, given the strains and stresses to which it will be subject.

— *Nanoprobe materials*: These are the materials required for the manufacture of tips for scanning probe microscopes, the basic tools of nanotechnology. These need to be chemically inert, physically stable materials capable of being fashioned reproducibly into atomic sharp tips.

— *Biosensors and transducers*: The capability of synthesising ultra high purity specialist organic chemicals having a range of terminating groups for these applications is required, as well as ways of bonding these molecules reproducibly to the surfaces of semiconductors and oxide materials

Advanced Manufacturing Processes

Manufacturing processes at the nanoscale can involve accretion or removal of material, or changes to the shape or form of material already present. Each of these processes provides new challenges and opportunities, as follows

— *Accretion of powders*: New generations of processing equipment will be needed to deal with nanopowders in the manufacture of nanocrystalline materials.

— *Quantum structures and devices*: The problem of producing devices with critical dimensions below 100nm, using 'top-down' techniques, is one that the electronics industry is currently wrestling with. Currently, commercial lithography is based on optical methods, but the wavelengths of visible and near ultraviolet light are too long to be usable on the nanometre scale. A range of alternatives is available, but parallel rather than serial writing techniques are needed for scale-up to commercial manufacturing levels, though this may not be a realistic goal for European companies.

— *Deposition*: Recent breakthroughs in the UK are making deposition on selected areas possible, in high transmission mode. Until now, this has been achieved only through focused ion beam sources operated in droplet mode - an approach which is restrictive in terms of the range of materials that can be handled.

— *Cutting, milling*: Only focused ion beam (FIB) techniques provide a means for selective cutting or removal of material with sub-100nm accuracy.

Although these techniques were largely pioneered in Europe - and the UK in particular - the present suppliers of such equipment are almost exclusively American or Japanese companies.

Machine Tools and Instrumentation

As structures become ever smaller, the necessity for on-line quality assurance test systems for certification duties, becomes more important and demanding. In the future, the nanometre scale will be the precision standard for material analysis, control purposes and also for material treatment. Already nano-analytical methods are used routinely for testing in the manufacture of:

— Magnetic storage disks

— Electronic multilayer systems

— Industrial polishing processes

New magnetoresistive multilayer systems offer drastically better positioning and controlling properties of sensors for application in the automotive industry and as measuring systems for velocity, strain or work piece positioning

Key areas of instrumentation and characterisation include

— Scanning probe techniques: observation + operation

— Some aspects of electron microscopy

— Some aspects of surface analysis

— Field emission + field ion microscopy + atomic probe analysis

— Nanomanipulators - using principles of mechanical/ optical/electric/magnetic/piezo techniques

— Test, calibration and measurement: Standards, benchmarks, procedures

— Nanotools, nanomotors, nanomachines

— Nanoprobes: production, characterisation, multiprobes

— Equipment to characterise magnetic/optical/electrical/ mechanical properties of nanostructures with high spatial resolution

— Microfluidics

— Focussed ion beam technology

— Computer software for data analysis and representation, simulation, modelling

An essential stage in the development of a large scale nanotechnology industry is the creation of machine tools for the production of nanodevices, and test, measurement and inspection techniques to aid manufacturers and provide quality control of nano products. It is also an area where knowledge has been, and continues to be transferred from research institutions to industry. It is the first nano area to become economically active, including the creation on numerous small and medium-sized companies.

Machine tools for nanotechnology are already being developed in Japan, the USA and (to a limited extent) in the UK. Not only is it already an economically viable aspect of nanotechnology, but its strategic significance is very high. These machines will have to underpin all future production of nanodevices, and it is very important that the UK should play a key role in their development.

Ultra-clean manufacturing facilities

Some aspects of nanoscale manufacturing may require clean room technology - either full scale facilities or 'table top' scale; but this will depend on the particular process or industry.

Industry needs suitable production methods for low cost manufacture of a whole range of materials such as nanomaterials, nanoporous systems, corrosion inhibitors, polymers, molecular sieves, ceramics, light absorbers and emitters, magnetic nanomaterials, pigments, colloids and so on. For end products, like catalysts or adhesive layers, a competitive market position can only be maintained if the analytical equipment necessary for material characterisation on an atomic or molecular level is available. Also essential to the equation are people who are trained to understand the new production methods, tools, analytical and testing techniques.

As materials at the nanometre scale may have unpredictable effects on living matter, the possible toxic and other hazardous properties of various nanomaterials need careful and sensitive investigation.

Sensors

Sensors are used to measure and monitor changes in every attribute of the physical fabric that affects our lives. If, as suggested by Foresight, cheap, ubiquitous, high-performance sensors are going to shape the coming decade, then nanotechnology will be the fundamental enabling technology for creating these products. Sensors have applications in every industrial process, in environmental monitoring, in defence, in the detection of toxic gases and in a range of 'electronic noses' with applications in industries as diverse as baking, brewing and medicine.

It is likely that the car industry will be one of the earliest adopters of new sensor technologies, sensing everything from emissions to predicting the imminence of impact. Goodyear has already announced their 'smart tyre' - one which uses microelectromechanical systems (MEMS), incorporated in the tyre wall. It will tell drivers when the tyre needs replacing or inflating. According to Goodyear, the sensors are cheap and reliable, and will remain stable for up to 10-12 years if required.

Car airbags provide a good example of how the cost of technology is decreasing, while functionality is increasing. In the USA airbags were once a major target for thieves because the accelerometer trigger they contain is so expensive. They are now becoming too cheap to steal, as an accelerometer on a single chip can be built for the equivalent of about $2.50, that is not only cheaper but also 'smarter' and more reliable. Future systems will incorporate sensors manufactured using nanotechniques that are capable of identifying not only the presence of a passenger, but their weight and size as well, and making adjustments accordingly.

Sensors presently utilise a range of technologies including MEMS (microelectromechanical systems), piezo-materials, micromachines and very large scale integration (VLSI) video:

— MEMS are created through machining, etching, cutting or milling to nanometre tolerances. Whatever applies now to the marketability and complexity of sensors based on MEMS will be hugely extended by these nanotechniques.

— Piezo-electric materials are materials that develop an electrical charge when deformed and, conversely, deform when in the presence of an electrical field. Piezo-electrics are particularly useful as surface-mount sensors for measuring physical movement and stress in materials. Like MEMS, piezo-electric materials have been around for some time, but new research is leading to the creation of new classes of piezo-based "smart materials" - materials that both actively sense and respond to the surrounding environment.

— Micro-machines are created using semiconductor manufacturing techniques, but are more complex in design than MEMS, incorporating in some instances micrometre-scale gears and other moving parts machined to nanometre-scale precision. Micromachines exploit the often overlooked structural qualities of silicon: a low coefficient of thermal expansion, high thermal conductivity, a strength-to-weight ratio more favourable than aluminum, and an elasticity comparable to that of steel.

Applications of ultra small machine technology are manifold - in microfabricated sensor arrays and neural networks; microelectronic noses; medical diagnostic breathalysers, and microfabricated terahertz electronic circuits and waveguides for medical imaging.

— VLSI Video. Today, a video camera with all the attendant circuitry required to attach it to a computer costs approximately £6 a unit to produce. This cost is dropping dramatically as all the components - the charge-coupled device (CCD), the circuitry and even the lens can now be packaged on a single chip which can be translated into a myriad of applications, including surveillance, security, and games.

The combination of cheap video and laser-based web bandwidth is already delivering a hint of what the future will really hold. The world of video is not a world of people looking at each other via videoconferencing. Rather, it is a world of cameras aimed at everything everywhere, watched over by machines, and only occasionally examined by people.

Global Positioning System Sensors

Global positioning system (GPS) sensors are being produced at lower and lower costs but with increasing performance. Systems once costing tens of thousands of £'s can now be bought in a handheld package for under £400. Nanotechnology means that in the not-too-distant future, integrated sensor/GPS modules will be small and inexpensive enough to integrate into packages in order to track the location and treatment of valuable goods they may contain. Laser technology is also rapidly changing gyroscopic technology as ring laser gyros (RLGs) displace traditional spinning-mass systems in aircraft, also delivering hugely increased performance and reliability at a lower cost. In the long run, it is likely that advanced MEMS accelerometer arrays will in turn displace RLG technology.

The impact of sensors will not stop at mere sensing. Sensing devices will be required also to respond to what they 'see'. The sensor era will then become a sensor/ effector era, where devices will not only observe the world around them, but will also be able to react to, manipulate and change what they 'see'.

VIRTUAL REALITY

Virtual reality offers us the means of observing and understanding how things work and enable us to interact with this new and previously unimaginable domain inhabited by molecules and atoms.

The term Virtual Reality (VR) is applied to a collection of technologies which together form an interface between humans and computers which is more intuitive to use. Typically (but not exclusively) this is achieved by real-time interaction with three-dimensional computer generated imagery (CGI). In some cases, exploring and interacting with the CGI requires the use of special peripheral devices to create a sense of 'presence' or 'immersion' within the virtual environment.

VR is a convergence of technologies from a much wider range of disciplines - human factors, telerobotics / telepresence, multimedia, computer-aided design (CAD), process simulation, ergonomics simulation (including mannequin tools) and computer-generated imagery (CGI - games, animation). One of the emerging strengths of VR is that it enables objects and their behaviour to be more accessible and understandable to the human user. This is particularly in cases where objects or processes exist in reality, but are not visible to the naked eye, as is the case with nanotechnology.

Using VR to visualise objects at micro-/ nano-scales is not a new idea. As long ago as 1992 VR was being used to visualise the output of a scanning tunnelling microscope (STM), effectively giving surface engineers the means to witness - and measure - atomic changes brought about by chemical and radioactive processes. The data produced by the STM are ideal for conversion into a VR model, which is typically made up of hundreds of individual polygons or facets. Instantaneous records from the location of a mechanical boom supporting the microscopes probe provide x and z coordinates of each surface point during the scan. Data relating to the y coordinates (surface relief) are obtained from tiny boom deflections, which occur as the system attempts to maintain a fixed current between the probe tip and the sample surface.

One company, Philips, has been evaluating Virtual Reality for some time, and late in 1996, the company used VR to 'miniaturise' employees so that they could experience the microscopic nature of the semiconductors they were employed to manufacture. This had been performed once before, using a miniature optical switching assembly. However, in the Philips example, the requirement was to miniaturise the users (virtually) to the level of the surface of the chip concerned. The demonstration starts when the VR user dons a head-mounted display equipped with a small electromagnetic tracking sensor which relays position and orientation information to the host computer, updating the images in the headset as the user's head turns. Once in the headset, the user finds herself in a simple office environment. The user is also provided with a handgrip, which is used to control her movements around the office. On a desk in front of the user is a printed circuit board, lying next to a PC.

As the VR user approaches the desk, she is shrunk to a height of around 1cm and is then free to move amongst the resistors, capacitors, microprocessors and the TrenchMOS semiconductors themselves. On approaching one of the semiconductors, the plastic cap automatically lifts off, exposing the TrenchMOS itself. The user is then shrunk again, this time to a height of around 1mm, and can move around the very reflective surface of the chip. The final shrinkage takes the user to a height of some 20-30 microns, and as the smooth aluminium substrate is removed, the regular hexagonal patterns of the chip surface are exposed, extending as far as the eye can see.

The implications of using VR for non-destructive testing and inspection of materials in the offshore, aerospace, electronics and medical industry are enormous, as is its use in assisting humans in the assembly of future micro-devices, such as motors, pumps and actuators. Add to this recent developments in the haptics interface arena (systems capable of providing the sense of touch and force using VR), and suddenly one begins to see the dawning of a new and exciting era in interface specifications for allowing micro- and nano-designers perform real design and assembly tasks.

Catalysis

Catalysis could arguably be the most important technology in our modern society since it enables the production of a wide range of materials and fuels, as well as being fundamental to the control of polluting emissions from industry and cars. However, the full extent of the nature of catalysis is not completely understood. Nanotechnology offers the key to a full understanding of catalytic processes, and when this happens, existing catalytic processes will be improved, and new processes designed and developed.

Most catalysts are 'heterogeneous catalysts', and are highly porous materials with a large surface area. Because catalytic reactions take place at the surface of the material, therefore the more surface area, the better. Many of these catalysts consist of an active phase and a passive phase, the latter being made of a chemically and thermally stable material (such as alumina). This support, or passive phase, enables the active phase to exist as nanoparticles that are resistant to sintering. The size range for such nanoparticles is from the atomically dispersed through to micron size, though commonly they are in the range of 1-10 nm. These materials were recognised long before it was realised they belonged to the realms of nanotechnology!

Areas of Major Impact of Catalysis

Fuel Production: Catalysis is crucial to reforming oil distillates to the appropriate chemical mix for use in cars, trucks, planes etc., (all fuels have been through several catalytic plants in their production). High levels of sulphur in fuels are removed to very low levels in some countries (many more in the future) by catalytic desulphurisation.

Materials Production: All polymers are produced using catalysis; some undergo a large number of different processes, for example, nylon production requires more than 12 different large-scale catalytic processes.

Environmental Protection: Catalysis is used in the clean-up of emissions from cars and industry, helping reduce atmospheric and aqueous pollution.

Health: Many drugs and pharmaceuticals are produced using catalysis. It could be said that chemistry, not medicine, drives modern healthcare!

Agriculture: Catalysis is used to produce fertilisers and crop protection chemicals.

Food: There are many applications here - for instance all margarine has passed through a Ni/SiO_2 catalyst to hydrogenate the double bonds on vegetable oil to give 'spreadability'.

Commodities: A huge range of basic chemicals is produced from oil and gas, using catalysis. One example is ethylene oxide used for antifreeze and for washing powder additives.

Properties of Nanoparticulate Catalysts

It is known that a variety of phenomena affect the reactivity of nanoparticulate catalysts. Pinning down the nature of the active sites has proved difficult and it is very rare to know even their surface morphology or how they work at the atomic level, partly because the techniques to enable such studies were not available. However, the advent of scanning probe microscopy (SPM) and other techniques are beginning to shed light on behaviour in this field at the nanoscale. A full understanding of the forces at work will enable better and more effective catalysts to be designed.

Coatings

Coating technology is now being strongly influenced by nanotechnology. Nano-scale materials, characterised by grain sizes of less than 100nm are being investigated, and significant advances are being made in the synthesis of high quality nanocrystalline powders, and several methods such as vapour condensation and solution precipitation have been scaled up for industrial use. Metallic stainless steel coatings sprayed using nano-crystalline powders have been shown to possess increased hardness when compared with conventional coatings, although the porosity is increased. Also, the reaction kinetics involving small scale particles during spraying are notably different. Novel tungsten carbide - cobalt coatings produced from nanostructured powders have also shown promising results in terms of bond strength, but to date little work has been conducted to assess their durability.

Nano-composite coatings comprising a ceramic/polymer mix have also been applied by High Velocity Oxy-Fuel (HVOF) spraying. These nanoparticle reinforced polymer composites have advantages in applications where the polymer strength is less important than the ductility of the coating. The HVOF process has also been shown to cause little polymer degradation. More studies are still required in this new area to improve the understanding of the interfacial properties at the polymer/ceramic junction, and how this affects the coating properties.

In addition to being able to apply coatings made from nano-phase powders, techniques themselves are being developed in which the processing parameters involved in the spraying actually produce the nano-crystalline structure. This has been experimentally achieved using a hypersonic plasma particle deposition (HPPD) process to apply SiC coatings. Coatings were successfully applied and the properties showed great promise.

Chemistry

Self Assembly and Self Organisation

What follows is a discussion of self-assembly in organic molecules, but it is worth mentioning briefly an inorganic example of self assembly - the Stranski-Krastonov method for growing of self assembly quantum dots (SAQD) which has rendered the lithographic approach to semiconductor quantum dot fabrication virtually obsolete. This method has been the catalyst of intense research, and UK groups are already working to develop SAQD lasers.

The study of the self-assembling nature of molecules, enabled by the tools of nanotechnology, is proving to be the foundation of an enormous explosion of applications in science and technology. The combination of being able to define structures on a 10 nm scale, produce nanoparticles with almost any desired radius, use new microscope technology and derived measurement techniques to 'see' surfaces and structures on a 0.1 nm length scale and also manipulate these structures, allows us to observe and harness new material properties that are a consequence of this self-assembly property

The physical properties of matter are highly size dependent. Given that for spherical particles in the nanometre range, the electronic energy level of separation is proportional to the square of the inverse radius of the particle, it is obvious that nanoparticles will exhibit many new and exciting phenomena that can be used in new applications.

In materials that in bulk form exhibit phase transitions phenomena associated with size will have major implications. The phase transition properties of, for example magnetic, superconductive and ferroelectric nanocomposites can be designed in such a way that devices can be created that operate just below those threshold fields or temperatures or currents required to switch the system from the collectively ordered into the disordered state. In this way, small perturbations can be made to give rise to large responses; and this feature can be incorporated in new types of detectors and sensors.

The behaviour of a small quantity of self-assembling molecules is strongly influenced by the way they are attached to a surface. Surface forces predominate in small or ultra thin assemblies, and in practice this means that the behavior of nano-assemblies can be more readily manipulated by the choice of surfaces.

It is even possible to envisage the possibility of programming self-assembling systems! A particular type of ordering may for example only be achieved if the system is subjected to a specific sequence of external signals and stresses. At present, any 'information' generated, in the form of a local change in structure, can be stored in organic and inorganic glasses, including self-assembling polymers, or by burning an electronic 'hole' in a molecule, using specific light signals. To create more advanced memories, it may be possible to harness the versatile organisational properties of small quantities of self-assembling molecular systems. Charged and dipole-carrying self-assembling systems have been shown to have a very large number of distinct configurations, some of which can only be achieved by specific and deliberate triggers. These can be manipulated to store, and perhaps later process intelligent information.

New Materials

It is also possible to make new types of nanocomposites or organic/inorganic hybrid structures by depositing or attaching organic molecules to ultra-small particles or ultra-thin man made layered structures (superlattices). There are many applications for these new materials in different products and processes, some of which are described below:

Application	*Improvements using new attached molecules*
Colloids:	New structural, optical and thermodynamic properties.
Pigments:	Better thermal stability without loss of absorption characteristics.

Dispersions:	Tailoring of nanoparticle size, and molecular structures.
Emulsions:	Engineering of viscosity and particle size.
Anti corrosion coatings:	Better surface mechanical properties, and stability in air.
Ferrofluids:	Creation of 'fluid magnets', which can be externally manipulated and be made to exhibit different rheological properties and structure.
Magnetic particles:	Below a critical size, magnetic particles are superparamagnetic and the collective magnetic response is fluid. A weak external magnetic field can produce a large response .
Passivated magnetic particles:	Magnetic particles can be covered with insulators or molecular overlayers, and the surface made resistant to oxidation or insulated to electrical conduction. These particles can be made bio-compatible and used in medical aplications.
Ceramic processing technology:	'Customised' ceramics for one-off applications; particularly using stabilised zirconia
Nano-emulsion:	Nanoparticle size and composition selected to produce required viscosity and absorption characteristics
Lubricants:	Tailored viscosity and thermal expansion properties.
Drug delivery systems:	Nanoparticles can determine the chemical reactivity rate, the location and the timing of drug delivery.
Bio-receptors for energy transfer:	Nanocomposites can be designed to exhibit well defined singlet or triplet excitonic absorption spectra. These localised bundles of light energy can be transferred over long distances to a 'receptor' and used for example for photochemistry or charge generation (e.g. the photoelectric cell).
Cosmetics: and UV protection gels	The addition of nanoparticles can influence the flow characteristics and mechanical properties of cosmetics, as well as the absorption of harmful radiation.
Sol-gel technology:	In the design of different types of materials.

Finally, nanoparticles can be alloyed to engineer surface bonding properties, and used for the fabrication of sticking, non sticking or selectively sticking surfaces.

New Devices

New depositional techniques offer a myriad of industrial applications, from the production of new chemical and gas sensors, optical sensors, solar panels and other energy conversion devices to bio-implants and in-vivo monitoring.

The basis of these techniques is an organic film (the responsive layer) which can be deposited on a hard, active electronic chip substrate. The solid state chip receives signals from the organic overlayer as it reacts to changes in its environment, and processes them. The innovatory techniques involved are

— making of the hybrid interface

— addressing the organic film from below (i.e. from the substrate side) using state-of-the-art superlattice and nano-engineering technology.

Some applications are as follows:

— Molecular films can be combined with optical waveguides and resonators to make evanescent field sensing devices with applications in biosensing and optical switching devices

— Molecules and bio molecules can be used to absorb and transfer photonic energy to chemical energy; with applications for solar energy conversion and photochemistry

— Surfaces of liquid crystals or thin membranes and other organic compounds can be used to detect molecules via structural, dielectric or conductive changes, with applications in gas and chemical sensing

— Semiconducting (or metallic) circuits and tunnel devices can be fabricated incorporating organic or biological materials with applications for imaging biomolecules and molecular motion, chemical reactions, growth, local temperature fluctuations and environmental changes.

Nanometrology

Fundamental to commercial nanotechnology is repeatability, and fundamental to repeatability is measurement. Many of the processes and production procedures envisaged in nanotechnology take place at the atomic level. In order to achieve adequate control of the movement of systems that might be used in such processes, for example for atomic manipulation, dimensions need to be known at the atomic level and beyond.

Moreover, a knowledge of the perfection of the texture at the nanometre and sub-nanometre level is an essential requirement if highly specialised applications of nanotechnology are to operate correctly, for example x-ray optical components and mirrors used in laser gyroscopes. Thus dimensional metrology plays a key role in the success and efficient implementation of many developments in nanotechnology. Ultimately all the measurements of length must be referred back to the national standards of length which are held at the National Physical Laboratory in Teddington.

How can the accuracy of position, size and displacement be achieved at the atomic level? There are basically two methods: First, by using a laser interferometer as an ultra-sensitive displacement transducer and sub-dividing the fringes to an extremely fine level. In this way displacements of the order of fractions of a nanometre can be measured which can be traced to the national standard of length. However, one problem with laser interferometer-based systems is that the components of the interferometer are not completely perfect and do introduce small uncertainties in the way that the optical fringes are sub-divided. Nevertheless, such interferometers can be used for measuring over an extremely wide range (from nanometres to several metres) and are therefore extremely valuable measuring tools.

The problem in the uncertainties of sub-division of optical fringes can be overcome by using an interferometer using x-rays. Using the diffractive properties of x-rays in the silicon crystal an interferometer can be constructed that has a fringe spacing of about 0.2 nm. Because much work has been undertaken to determine the lattice spacing in silicon with exceptional accuracy, this method provides a very powerful method for measuring ultra-small displacements. But such interferometers can only perform measurements very slowly.

In order to meet the requirements of visualising surfaces at the atomic level a wide range of probes are capable of revealing the positions of atoms. Whilst, such probes can generate very impressive pictures of atoms on a surface, however, it is only relatively recently that such probing systems have been able to be used for measurement. By fitting laser interferometers to the microscope the tip of the microscope can be tracked accurately in space. NPL has a system that can measure components, mostly surface texture reference standards, to an accuracy in the order of nanometres.

ENERGY

Nanotechnology has an important part to play in the generation of renewable energy particularly through the enhancement of photovoltaic (PV) cell technology. The conversion of sunlight directly to electricity offers an inexhaustible and environmentally benign energy source. If the efficiency of PV cells were enhanced and costs reduced, they would make a very significant contribution to the energy needs of the developed and developing world.

Current concerns over CO2 emissions are presently driving an increased interest in PV. Although total PV power output remains relatively low, the industry is growing rapidly - the production of PV modules expanded 40% worldwide in 1997. Although most current PV production is based upon crystalline and amorphous silicon technologies, research is now focusing upon new technologies which may result in significant reductions in PV costs, and / or improvements in efficiency. The two areas in particular where nanotechnology is playing a key role in achieving these aims are:

Quantum Well Solar Cells

The cost of PV devices can be reduced directly by the use of light concentrating systems, and indirectly by using narrow-band gap cells to convert waste radiant energy from conventional power sources. The latter is a novel, fast growing area known as thermophotovoltaics (TPV).

The incorporation of quantum wells (QW's) is being studied to produce higher efficiency photovoltaic cells for concentrator and TPV applications. QW's are ultra-thin layers (nanostructures) of narrower band-gap semiconductor (the well) grown between regions of higher band-gap material (the barrier) by modern crystal growth technologies such as molecular beam epitaxy (MBE) and metal-organic vapour phase epitaxy (MOVPE). These techniques are already being extensively researched for information technology applications. In photo-diodes and modulators they will be the basis of the all-optical computer of the future.

Dye Sensitised Nanocrystalline Devices

The second example of nano-structured materials for PV applications concerns the use of nanoporous metal oxide films. In 1991, Michael Grätzel developed a novel photoelectric cell that closely mimicked the principles of photosynthesis. Light is captured by an intensely coloured dye; the energy of the light is then used to inject an electron from the dye into an semiconductor such as titanium dioxide (TiO_2). This remarkably efficient charge separation reaction initiates current flow and the output of electrical energy by the cell.

These novel photoelectrical solar cells have attracted considerable interest because of their potential as low cost solar cells. In contrast to conventional solid state solar cells, which depend on advanced processing of very high purity semiconductors and are therefore relatively expensive to make, the Grätzel solar cell is fabricated from cheap, low purity materials by simple and low cost procedures. There is therefore considerable excitement at the possibility that development of these novel devices will result in cheap and environmentally friendly energy production.

Preparation of these cells depends critically on the type of semiconductor used. The ideal material must have a high surface area for light absorption and charge separation. In the Grätzel cell, a high surface area is readily achieved by using nanometric particles of TiO_2 (diameters 10-20 nm). For fabrication of a photoelectrical cell, these nano-dispersed TiO2 particles are sintered together to give a high surface area, nanoporous thin film supported upon a glass substrate. In keeping with the simple fabrication procedures involved with these films, this procedure is based upon silk screen printing techniques from a colloidal suspension of the TiO_2 particles, followed by heating the film using a hot air blower. The resulting films combine high surface area with optical transparency, excellent stability and good electrical conductivity.

The remarkable range of properties of these films have practical applications, not only for solar energy conversion but also for photocatalytic water treatment, photochromic windows and molecular electronic devices and encapsulate the essence of the opportunities offered by nanotechnology.

MATERIALS

The world market for materials is estimated at $10 billion p.a. and growing. Since the 1920's scientists have known that the properties of materials such as strength and the ability to conduct electricity were governed by the structure of their atoms and molecules. This insight led to the identification of semiconducting materials that laid the foundations of today's electronics industry.

More recently, with the advent of the tools of nanotechnology, materials science has been transformed to a point that the relationship between the structure of a material and its properties may be controlled. Scientists and engineers are becoming increasingly able to understand, intervene and rearrange the atomic and molecular structure of matter, and control its form in order to achieve specific aims. Materials are closer to being designed to fit the needs of a specific end use, and simulate the properties of specific materials even before they are made.

This ability is not the result of a continuous evolution of knowledge, but a step-change resulting from increased computer power (itself a result of progress in materials science) and advanced instrumentation, which allows complex mathematical modelling of materials from the micro-structural to the quantum mechanics level. This modelling relates the properties of a material to its internal structure, and even its behaviour when processed. The properties and performance of a material can now be tailor-made, and even information on the behaviour of a material that has been modified by advanced surface treatment, coating, joining and adhesive technologies can be designed into the manufacturing process itself.

Computer modelling and simulation at the atomic level is already being used to improve the performance of currently available materials. Industry will eventually be able to design entirely new materials, and build them atom-by-atom and molecule by molecule. The arrival of such nanophase materials at the commercial scale will be accompanied by new processing and fabrication technologies. From this, nano net shape devices such as sensors, robots, electronics systems, computers, engines and surgical devices can be produced with applications in medicine, pharmacology, agriculture, mining, genetic engineering, energy and the environment. These developments will also dramatically cut the R&D and fabrication cycle times.

The consequences of these revolutionary developments in science and engineering need to be addressed from a long-term perspective both by firms and nationally. There are cumulative gains to be had from the capability to design, produce and use new materials in advanced nanotechnology-based products. Time is short - it may soon be impossible for new companies not already involved to enter into the field of advanced materials.

8

Nanotechnology and Business World

The history of predictions about the societal and economic impacts of promising new technologies is replete with predictions that incipient advances will amount to little, only to have them substantially transform daily life, and those about major advances that subsequently fizzle.

Recognizing that inherent in forecasts of revolutions is the premise that "things will be different this time," this stance leads toward reliance on the research literature to construct an analytical framework. Economists, as well as scholars in other fields, have long studied the generation, diffusion, and impacts of scientific and technological innovation. Findings from this body of research do not themselves constitute predictions about the future of nanoscience and nanotechnology, but they do outline the variableslikely to determine the rate and direction of these impacts and to identify relevant research questions.

What follows is a brief distillation of mainstream propositions about the economics of scientific discovery and technological innovation, customized to nanoscience and nanotechnology. Particular emphasis is placed on what have recently been labeled "general purpose technologies," — that is, enabling technologies that open up new opportunities across a swathe of economic sectors and production relationships. Although not explicitly articulated as such, each of these propositions constitutes a researchable question about the direction of public sector support for nanoscience and the rate and direction of the technical development and commercial deployment of nanotechnology.

Nanoscience and nanotechnology provide an excellent testbed case to study increasingly commonplace statements about the blurring of distinctions between science and technology and the speed at which new scientific findings are transformed into commercially important technological innovations. Current developments at the frontiers of research in these domains also provide a natural experiment to assess alternative models of relationships between scientific and technological advances.

The promise of nanotechnology is its ability to do some things that cannot be done by current technologies and to do some things "better" than are provided for by existing and latent technologies. Better relates to technical performance: smaller, faster, stronger, safer, reliable, even cheaper. The projected impacts of nanotechnology writ large thus represent the summed demand for these performance enhancements across the several potential uses claimed for it. Demand curve characteristics relate to projections about thevalues assigned by consumers to the specific performance attributes of the new technology relative to its price and that of alternative technologies.

The projected uses of nanoscience and nanotechnology encompass public sector (e.g., space travel) and private sector (e.g., manufacturing) final demands. Differences in the characteristics of demand schedules between the two sectors may affect the direction of scientific research and innovative endeavors. Public sector demand, for example, may be expected to emphasize performance characteristics, as in defense and space research, and thus to be highly price inelastic. One might thus expect the early uses of nanotechnology to be in those products produced to meet public sector demand. How these projections of market entry in turn feed back upon the posing and priorities of scientific questions is another analytical and policy-relevant question that warrants study.

In terms of the private sector, an initial conjecture is that nanotechnology goods and services will be introduced into those markets where performance characteristics dominate alternative techniques and for which demand is highly price inelastic. Relatedly, one would expect that the impacts of nanotechnology would be longer delayed and possibly less dominant in those markets where alternative techniques were competitive on a performance basis, and consequently, where demand was more price elastic.

Nanotechnology is presented as having potentially pervasive benefits — materials and manufacturing, nanoelectronics and computing, medicine and health — to cite but a few of the areas of use from the National Technology Initiative report and Drexler's *Engines of Creation*. Pervasive relates to possibilities; not all possibilities are equally profitable, however. The timing and sequencing of where nanotechnology's impacts are felt, in part, is a market phenomenon. Different cost considerations enter into making nanotechnology technically and commercially viable for different end users. As demonstrated in Griliches's classic study on the adoption and diffusion of hybrid corn , suppliers rationally order the sequences of markets for which they customize a general purpose technology.

Forecasts of the puissant impacts of new technologies often are voiced in tones of Schumpeter's metaphor of creative destruction — gale-force events in which a new technology quickly and completely replaces earlier "obsolete" techniques. In fact, the process of displacement can be both slow and incomplete, with use of earlier technologies continuing to be economically rational for extended periods rather than indicative of conservative or laggard behavior.

Diffusion of the new technology, and thus the range and magnitude of its societal impacts, represents displacement. In part, displacement is propelled by continuous advances in the new technology that expand thetechnical range of its use and possibly lower its relative price. In part, though, the process is slowed by renewed attention to improving the range of uses of earlier technologies and/or relative price reductions in this technology that preserve for it markets in which price rather than performance is a dominant selection criterion.

Perhaps the greatest difficulty in predicting the societal impacts of new technologies has to do with the fact that once the technical and commercial feasibility of an innovation is demonstrated, subsequent developments may be as much in the hands of users as in those of the innovators. Consider how the Internet has rapidly progressed from a technology supported by ARPA to facilitate communications among universities with ARPA contracts and to experiment with digital communications systems, to a means by which teenagers and college students exchange music files. In the process, the societal issues have changed from involvement of universities in defense-related research to legal suits over intellectual property rights.

The diffusion and impact of technological innovations is often as much a function of the development of complementary technologies and of a network of users as it is of the introduction of the discrete technology. The impact of railroads on economic growth relates to the iterative pressures that development in each component — engines, rails, brakes, signals, organizational structure — placed on the other; the value of the telephone or Internet is a function of the number of users connected to the system as well as to the number of computers that can be accessed for information. The technical or economic specification of either complementarities or network economies for nanotechnology appears to be at an early stage.

To paraphrase Paul David, the shift from one technological regime to another is a journey, not an arrival. To further quote from his "Computer and Dynamo," "many intricate societal and institutional adjustments, transcending in complexity and uncertainty the redirection of private investment planing, are usually entailed in effecting the passage from one "technological regime" to another.

On this view there are likely to be many difficulties and obstacles that normal market processes cannot readily overcome." What appears to be growing consumer resistance to genetically modified foods should serve as a cautionary example that the world does not necessarily beat a door to those who build a better mousetrap, however small it is. What needs to be better understood are the reasons for acceptance, resistance and rejection, both economic and non-economic.

Nanotechnology and Future of Business

The rapid evolution of advanced technology has constantly served up innovation after innovation in super-compressed time frames — from the mapping of the Human Genome and cloning to supercomputers and the Internet. This is an astounding metric validating we are entering an era driven by accelerated technology developments, that have increasingly a significant economic value. The rapid advance of new technology has moved beyond our ability to accurately forecast with precision the impact on economics, business and society. We need to approach this challenge with new predictive models that are designed for the real-time complex changes that emerging technologies are influencing. This is perhaps most relevant given the challenges of nanotechnology.

We are in the midst of a large-system paradigm shift driven by accelerated exponential growth of new technology. We are witnesses to faster, more comprehensive change shaped by new technology than any civilization in history. This is but the beginning of a new wave of technologies, such as nanotechnology, that will redefine, reshape and eventually transform economies and societies on a global scale.

Nanotechnology may become an essential large-systems strategic competency that will require coordination among all sectors of society in order to become a force for enhanced social productivity. This technology is fast emerging. Nanotechnology may well shape the sustainability and wealth of nations, organizations and entire industries in the future. A central concern here is the necessity for us, together as a nation, to plan today to meet the readiness challenges that most certainly will lie ahead.

If nanotechnology, the manipulation of matter at the atomic level, at maturity achieves even a fraction of its promise, it will force the reassessment of global markets and economies and industries on a scale never experienced before in human history. The ubiquitous nature of nanotechnology as a fundamental design science will have applications for numerous industries: manufacturing, health care, and transportation to name a few. Since we do not know what yet is possible we can only speculate on the potential. Those societies and interests that develop the next generation tools will be first to building the nanoeconomy of the 21st century.

Nanotechnology and Global Competitiveness

Recent developments in emerging technology and its impact on business and economics would indicate that forecasts are less than accurate in predicting the future. Few would have accurately forecast innovations such as of the Internet, wireless communications or the mapping of the Human Genome. Also, there have been numerous wild forecasts that have historically seemed more like science fiction than fact. Predictions about nanotechnology have fueled the imagination. Much of this is still imagination but the future looks promising. Nevertheless, new innovations in technology are reshaping the global economy at a dizzying speed.

It would be prudent to consider the possible economic outcomes given the accelerated emergence of advanced technology. To not be prepared, to spurn readiness would be unwise given the promise of nanotechnology. It is with this in mind that we turn to nanotechnology. Why is the potential economic impact of nanotechnology so important to consider?

Nanotechnology is a fundamental design science, yet to emerge, mostly theoretical today, that may well provide us with the tools to engineer inorganic and organic matter at the atomic level. Nanotechnology, if even partially realized, over the next few decades has the potential to realign society, change business and affect economics at the structural level. New business models, design tools and manufacturing strategies may emerge at price points much reduced and highly efficient.

Nanotechnology will touch all aspects of economics: wages, employment, purchasing, pricing, capital, exchange rates, currencies, markets, supply and demand. Nanotechnology may well drive economic prosperity or at the least be an enabling factor in shaping productivity and global competitiveness. Again, we are free to speculate in the dawn of such a new science.If developments in nanotechnology reach a critical mass in supplying radically innovative breakthroughs in automated self-assembly, as one example, most vertical industries will be influenced. Most industrial and post-industrial supply chains will be changed.

What if the fabrication lines for making computers are reduced in costs by 50%? What if drug development and manufacturing costs are reduced by 70%? What if energy sources were not dependent upon fossil fuels? What then might the impact be if nanotechnology were applied to real cost reductions for essential goods and services that affect quality of life, health, habitat and transportation? There would be a dramatic impact on lifestyles, jobs, and economics. Most value chains, supportive linkages, alliances and channels of distribution will be altered. Institutions of learning, financial services and certainly manufacturing will be reshaped.

We must learn to ask the questions now about how nanotechnology may change our choices, affect our lifestyles, shape our careers, influence our communities — we must ask now and prepare so we may examine the implications that may shape the future we will live in together.

The issues that remain are to consider in what timeline what actions might be taken. How might we prepare as a society for these changes? Will there be radical dislocations or a smooth coordinated adaptation? We must plan for multiple scenarios. Radical nanotechnology innovations potentially unleashed on immature markets, fragile economies and a business community ill prepared for rapid post-industrial transformation would be problematic. We see today alterations driven by e-business and the Internet already causing deep change to industries and economies worldwide.

Imagine the emergence of a nanochip that tomorrow would deliver over 50 gigahertz of speed with the processing power of ten supercomputers for the price of a quartz watch and smaller than a key chain. What might the economic impact on the computer industry be overnight?

Imagine a super-strong and inexpensive material to be used for construction and manufacturing that would eliminate the market for steel and plastics. How might that influence the economy?

In a world being reshaped daily by innovations, the absurd today is reality tomorrow. But with the intimate inter-linkage of markets, industries and economies radical breakthrough technologies will have a widespread and far reaching impact — positive and negative. It is entirely possible that, just as computers and the Internet have become vital linchpins woven into the fundamental economic landscape of today's strong economy, nanotechnology will emerge as one of the key technologies that shapes the future economy. Many of the necessary factors are in place to drive this scenario: widespread potential cross-industry applications; fast track R&D; government investment. The risks in not preparing for and examining the economic and business impact are too large to ignore.

In an era of prosperity it is difficult to consider the lack of global leadership that might befall a nation such as the United States. How might the United Kingdom have better prepared for its 19th century challenges if it had known what was to come at the height ofits global leadership in the last century? We might well ask the same questions today.Readiness is always a wise choice. Especially when it appears we may not need to be vigilant. Nations today — ill prepared to capitalize on the Internet, the transformation of supply chains or the mobile commerce sparked by advanced telecommunications — are playing catch up and it has hampered their productivity, GDP and competitiveness.

In 1999 the Institute for Global Futures deployed a privately funded study to assess the general awareness and readiness of the business community regarding the economic and business impact of nanotechnology. A series of interviews with a broad range of business executives in health care, manufacturing, medicine, real estate, information technology, consumer goods, entertainment and financial services was conducted, and is still being conducted at this time.

The Institute for Global Futures, a ten-year-old San Francisco organization advises the Fortune 1000 and government on the impact of leading-edge technology on markets, society, customers and the economy. The Institute covers telecommunications, robotics, computers, life sciences, the Internet, software, artificial intelligence and a host of other technologies and forecasts trends.

Overall, the level of awareness and readiness is low, based on the survey results. Less than 2% indicated that they thought they knew what nanotechnology was. An additional 2% had heard of nanotechnology but could not explain what it meant. Of those surveyed, 80% agreed when nanotechnology was explained in basic terminology that this was an important technology that had the potential to affect them and their business; 45% expressed an interest in learning more about nanotechnology.

Though one could question at this time, when nanotechnology is still in its infancy, largely theoretical, why should anyone care and why would we even expect readiness? The issue is one of accelerated change and its impact on business and society. We are interested in readiness and awareness prior to the accelerated changes that may lie ahead, and not so far ahead as we might think. Important issues regarding research and development in nanotechnology are present today. There are real issues that bear examination today as we plan for the impact on tomorrow. Readiness is central to adaptation.

Readiness regarding education, capital, talent, coordination, and communications are all integrally part of the same platform. As nanotechnology may translate into the sustainability of nations, organizations and entire industries —readiness, the preparation and planning process, becomes vitally important to define and examine.

Scenario One: Brave New World

Economic Environment: Nanotechnology comprehensively integrated into the economy due to high readiness, effective strategic planning and widespread investments by business, education, labor and government. Accelerated national policy and investments producing economic agility and rapid widespread large system change management. There is a widespread understanding of the numerous benefits from applications of nanotechnology, its strategic economic value for the nation, and its role in maintaining global U.S. leadership. Comprehensive social and industry-wide adoption has led to a positive impact on national productivity and an enhanced quality of life.

Key Characteristics: Robust gross national product; high productivity; global trade leadership; sustainable economic growth; global patent leadership; superior industrial competitiveness; integrated education and training resources; strong investment climate; plentiful capital liquidity; high investment on R&D; low unemployment; high government and industry collaboration.

Future Outlook: An ever-escalating predominance in key markets and industries leading to increased investments and innovations. An accelerated progressive and confident growth prognosis for the economy, and an enhanced quality of life for the nation. Global leadership and empowerment of third world and developing nations increasing. Accelerated investment in R&D and continued coordination with all sectors of society.

Scenario Two: Playing Catch-up

Economic Environment: Nanotechnology partially integrated into the economy due to low readiness and inadequate strategic planning. Economy playing catch-up. Slow social and industry-wide nanotechnology adoption. Reactive cultural reaction to investment and organizational and industry leadership for accelerated national change management. Not a full commitment and investment in national nanotechnology policy.

Key Characteristics: Partial loss of leadership in key markets and industries; Lack of skilled talent; poor education and training; growing but still low investment in R&D; fragmented industry support; poor investment climate; liquidity insufficient; fragmented government and industry collaboration.

Outlook: Optimistic if rapid and strategic widespread large-systems change is undertaken in a concerted effort by business and government partnership. Difficult to regain ground in certain markets, but partial leadership in key markets is a success to be built on for the future.

Scenario Three: The Bumpy Road (

Economic Environment: Absence of comprehensive nanotechnology integration, adoption and readiness leading to a drastic reduction in post-industrial growth, poor performance in global competitiveness with a negative growth impact on the overall economy. Denial of the strategic value and importance. Inability to invest in the actions required to manage comprehensive large-system socio-economic change.

Key Characteristics: Loss of key markets and industries; rising unemployment; chaos in selected sectors; brain drain going offshore; lack of investment liquidity; low investment in R&D; fragmented business and government collaboration; flight capital moving offshore; educational support low.

Outlook: Moving forward into the future, it will be difficult to seize and attain market and industry leadership without a significant investment in R&D, education, training and private/government collaboration. A commanding market share in key industries and global leadership will have been sacrificed. Regaining this ground, certainly global leadership, will be a massive undertaking certain to strain capital and human resources. An acceptance of a less involved global leadership role will be the probable outcome.

Nanoeconomy and Future Wealth of Nations

As the global economy continues to be transformed by new technology, a keen competition will develop for talent, intellectual property, capital and technical expertise. We see many of these factors responsible for shaping how nations today compete, interact and trade. Technical innovations will increasingly shape economies and market robustness.

Technology will continue to drive global and domestic GDP. Competition will be fueled increasingly by fast breaking innovations in technology. Today this is obvious as rapid technological changes in telecommunications, life sciences, and the Internet demonstrates the emergence of entirely new economic and business realities. If the proliferation of today's technologies to form new business models is any indication of the speed and power of change in the economy, future nanotechnologies will make for an even more dramatic paradigm shift.

The evolution of a nano-economy, as contrasted with the petro-economy of today, is an intriguing idea. How might an economy not dependent on oil realign itself? More study will be need to be conducted in order to understand and map these scenarios.

Fundamental nanotechnology innovations yet to come will set the timeline for this economic transformation. Or, nanotechnology may just become integrated into industries such as health care, manufacturing and energy much like artificial intelligence became an embedded component of new products.In conclusion, the readiness of a nation to prepare for large-scale economic change is a challenging task. Nevertheless, the future wealth of nations, certainly the economic sustainability of nations, will be shaped by the preparations we make today.

Coordinated large-systems strategic planning efforts may well shape our ability to adapt. Strategically important decisions will need to be made. Vastly important national security and economic issues lay yet unexamined. Huge cultural issues related to managing large-scale change will need to be better understood and plans formulated.

Nanotechnology provides a stimulating and somewhat awesome challenge to meet. If we had the knowledge in the 1960s and 1970s to prepare for the impact of computers or telecom in the 1990s, how might we have prepared the nation? Today we have real-time examples and a history of rapid accelerated economic change due to new technology to learn from, in preparing for the future. It is too trite to state no one can know the future. The future may indeed be unpredictable. But we do know that without asking the hard questions, without speculating on the possibilities, without preparing the nation by building readiness, we may do ourselves a disservice that will be difficult to repair.

As nanotechnology moves from the theoretical to the practical, as many of us believe it shall do faster than is expected, then the possible impact on business, society and the economy will become evident over time. But we have a new opportunity today. Given the recent history of digital technology and access to better models of socio-economic analysis, we must consider growing readiness a social responsibility. We must consider readiness as part of our social policy.

We might well consider the possible futures that will result from our collective actions. We must have the courage to speculate on the possible nanotech futures we may shape as a nation. This will determine whether we have a Brave New World or a Bumpy Road.

Making a Science Policy

The physical sciences are attempting to find a new basis for public funding, recognizing their relative eclipse by the biological sciences. The biological sciences are publicly funded on the basis of explicit or implicit promises to cure diseases although creation of new industries and jobs has also become a theme in budget negotiations in recent years.

A social needs based strategy potentially raises the profile of the social sciences in formulating and implementing science policy. A social needs approach also brings engineering to the forefront and changes its relationship to basic research by reversing thedirection of the linear model. Instead of going solely from the serendipitous results of basic research to utilization, with engineering playing the role of operationalizer, in an alternative model, engineering expertise plays the role of translating social needs into broad targets for basic research.

The social sciences are also brought into the picture to help formulate the social needs that are expected to be the basis for the development of new scientific disciplines as well as to identify unintended negative consequences of new technologies. The emerging strategy for NSF thus brings together the three elements that have often had an uneasy co-existence in the same organizational framework into a new cooperative relationship. In the following we discuss some of the implications of a social needs based model of science policy for nanoscience and the social sciences.

Alternative Science Policy Models

Two fundamental models of science policy were adumbrated during the early post-World War II era:

(1) the run-off of useful results from the "meandering stream of basic research." This became known as the linear model which presumed a one-way flow from basic research to practical results, and;

(2) a zigzag line running from identification of social needs and technologies designed to implement a solution.

This latter model was expected to be supported by directed basic research targeted at technological goals with fundamental results a serendipitous outcome. Both models can be found in the 1945 Bush Report, *Science: The Endless Frontier,* although in subsequent years the basic research modality largely crowded out the social needs dimension of science policy.

The technological outcomes of basic research assumed the forefront of attention due to the tremendous power of a single instance. Lise Meitner's findings, in the mid-1930s, that suggested fissionability of the nucleus of the atom became the exemplar of the linear model of basic scientific research leading to practical applications. The physical sciences achieved pre-eminent status and a public funding rationale during the early post-war through the success of the wartime Manhattan project. Other successes, such as radar, and the transistor in the early post-war period, based on other R&D models, also helped.

The alternative approach, "a reverse linear model" of placing social needs first, is not a new ideal. Spaceflight goals such as the Apollo Program are an instance of a social end inducing technological development work, as well as more fundamental research, both as a support structure and as an outcome of the technology project. As the agency with primary responsibility for the U.S. space mission, NASA has assumed responsibility not only for issues directly concerned with its mission, such as increasing university S&T capacities related to space, but also cognate issues such as improving science education and technology transfer.

Crisis of the Linear Model

Wartime success and the promise of peacetime spillovers justified the creation of Federal agencies such as the National Science Foundation and the Atomic Energy Commission, which have been mainstays of support for much of the physical sciences. The shock of the funding cutoff for the Superconducting Supercollider is the proximate cause for re-evaluation of future public funding prospects in the physical science community.

Attempts to renew funding based upon an assumption of a "contract" between science and society to provide funding in exchange for long term future unexpected benefits have not worked well in recent years. Such open-ended appeals have been met with skepticism. The original Bush model of the Endless Frontier was built upon a catalog of social needs, including housing and national security, which science was expected to address. However, with notable exceptions such as space and the abortive civilian nuclear energy initiative, the Endless Frontier model shifted from a focus on social needs to one of basic science as an end in itself, serendipitously spinning off unexpected useful technologies.

Private Sector Version of a Social Needs Model

Social needs, or at least those that pass the tests of economic viability and fit with company strategy, are mainstays of private sector science policy. In recent years, such "policy" has increasingly been formulated at an industry as well as individual firm level. With respect to nanotechnology, the search for a viable alternative to silicon, with its limits in computer chip design, is a strong impetus. Moreover, the relative balance between public and private science policy making is shifting, even as the two are increasingly formulated jointly.

Industrial R&D spending having passed private spending several years ago augurs a change in the relationship between public and private science policy. This secular shift has arguably changed the context for public R&D funding. Heretofore, public spending provided the base from which private R&D took off. The contemporary situation is mixed. The classic instance, the invention of the transistor at AT&T Bell Laboratories, was based on corporate support for fundamental research in solid state physics, which was targeted at improvements in telephone switching systems.

Private capital is regaining the pre-eminent status as a source of support for scientific research that it held in the pre-war era in tandem with private foundations. Private support is increasingly important for the biological as well as the physical sciences. Such support is increasingly allocated as venture capital to new firms rather than through corporate research budgets of established firms. The potential of nanotechnology to induce a wave of start-ups will allow academic researchers and their business partners to tap into capital markets for R&D support, following their predecessors in biotechnology.

Toward A Public Sector Social Needs Model

Nevertheless, public funding is still extremely important, especially as the basis for academic research. It is responsible for maintaining the traditional knowledge flow to industry through publication and consultation. It also provides the base to leverage private funds as well as foster the more recent and widespread creation of intellectualproperty, which ensues from the Bayh-Dole technology transfer regime. After decades of living off World War II successes, physical science leaders recognize that a new tack has to be taken to gain public support.

Technical advance by itself will not suffice to capture the public imagination. Touting the ability of nano devices to store the equivalent of the Library of Congress in an exceedingly small electronic space produces a "ho hum" reaction in Congress and among a public accustomed to amazing technological feats.

At the same time physical scientists realize that the success of the biological sciences has not been unalloyed. Significant opposition has arisen to technologies arising from biological research. It was mentioned at the meeting that some companies, heavily invested in biotechnology, attempted to inhibit discussion about genetically modified foods. The reaction against suppression of debate tends to leave the impression that there is something wrong with the technology itself, a conclusion which may or may not be warranted.

A more open information and discussion process was called for. Recognizing these deficits in the biological sciences model and the lessening effectiveness of their own model of seeking research support, the physical sciences seek to emulate the biological sciences while remedying the defects in the bio-science funding generation model.

The two-fold strategy is to find a basis in social need to justify physical science research, on the one hand, and ally with social scientists and ethicists to analyze and head off potential ill effects or misunderstandings of emerging technologies. Thus, for example, researchers at Sandia Laboratory have extrapolated an extension of their mission of national security to global security. Within that expanded purview they have identified the problem of potable drinking water as a target for nano-membranes.

Another theme is the physical sciences as the basis for future advance in the biological sciences. Just as physical science models drawn from bio-physics and chemistry became the basis of molecular biology and biotechnology in decades past, nanoscience is predicted to be the source of future fundamental discoveries in biology.

There is a three fold role for the social sciences in nano:

(1) analyzing and contributing to improvement of the processes of scientific discoveries which increasingly involve organizational issues where the social sciences have a long-term research and knowledge base;

(2) analyzing the effects of nanotechnology, whether positive or negative, expected and unintended, hypothetically and proactively and as they occur in real-time;

(3) evaluation of public and private programs to promote nanoscience and nanotechnology.

Given that future economic growth and societal advance are increasingly based on S&T, understanding the social processes by which new fields of science and technology are formed becomes more than an esoteric scholarly interest. The classic work of Joseph Ben-David on discipline formation in the 19th century needs to be renewed to take account of changed conditions. More recently, the formation of new scientific and technological fields appears to occur not only through extension and bifurcation of single fields but through a process of hybridization of elements of various fields, some of which had little or no previous interaction with each other.

The process of field and discipline formation requires investigation through in-depth interviews and bibliometrics. A comparative series of studies would be most useful to produce theoretical generalizations as well as practical advice on how to benchmark and induce the formation of new fields.

Policy Implications

Programs of networked research initiatives could also be deliberately constructed to induce new forms of collaborations in target areas along the lines of the European Union Framework Programs. The Canadian Network Centers of Excellence provides another model of large-scale collaboration, primarily among university research groups. The U.S. Advanced Technology Program emphasizes collaborations among firms, large and small,but academic researchers can also be included. Considerable social science based evaluation studies have been conducted on these programs which could be tapped as a basis for organizational design of new programs.

9

Future of Nanotechnology

The history of information technology has been a history of learning to make "bits" smaller. There is no obvious and hard physical limit to the minimum size of logical devices that process information or the marks that store information. Indeed, quantum physics is currently being recast as a theory of information, and even a single atom can no longer be seen as the ultimate limit to the minimum size of a bit. Moreover, the smallest logical devices being manufactured currently contain billions of atoms, and the smallest magnetic bits on commercial hard drives contain millions of atoms.

From a scientist's point of view, there is a long way to go in the development of information technology. At current rates of progress, logical devices based on silicon and information storage based on magnetism will reach physical limits to size reduction in perhaps ten to fifteen years. Unless entirely new logical devices and entirely new storage devices are invented and brought to manufacturing, important economic forces driving the development of nanotechnology will be weakened.

Another twist in the technical road will occur when lithographic manufacturing processes can no longer be extended to smaller dimensions and must be supplemented or replaced by processes of natural assembly as described above. Rather than carrying silicon microelectronics and magnetic storage relentlessly forward, such technologies may first be established in niche applications where the cost of the established technologies can be easily undercut. The manufactured structures may not be particularly small.

But let us ignore for the moment the likely twists in the technical road, and instead assume continued exponentially compounding improvement in our ability to build complex structures with ever-smaller critical dimensions. Worldwide competition in microelectronics will certainly drive us in that direction for at least the next decade, but even at the end of a decade, atoms will be small. An additional 25 years will be required to develop the capability to fully design and control the structure of an object on all length scales from the atomic to the macroscopic — in other words, the beginnings of a mature nanotechnology. Society has some time to cope with, react to, and learn to harness this capability.

To grasp some implications of a mature nanotechnology, imagine a world where information technology is truly ubiquitous and dirt cheap, where even trivial human artifacts contain extraordinary complexity and therefore extraordinary ability to store, process and communicate information. These broad capabilities of future information technology are easy to forecast, but their implications for society are still difficult to discern. History suggests that the most important future applications will surprise us. Who, among the visionaries, architects, and early developers of the Internet, predicted electronic commerce as its "killer app"? History suggests that the trivial applications will also surprise us. What a leap of faith it would have been, if the builders of ENIAC (the first vacuum tube computer) could have imagined that in fifty years the computational equivalents of ENIAC would be built into greeting cards, for the sole purpose of saluting each card recipient with a trite melody, each "ENIAC" to be discarded after twenty seconds of playing time.

Much public discourse currently centers on the ease with which copies of digital data can be made, what constitutes fair use of this capability, possible technical solutions to limit use, and the impact of all this on copyright and intellectual property law. Our present quandary over copyright law may seem quaint in a future where reproductions of any object are increasingly inexpensive and increasingly indistinguishable from the original.

The costs of digital information storage and communications bandwidth have dropped to the point where it is now feasible for a large population to copy, store, and transfer music digitally. Gurley points out what has just become feasible will rapidly become trivial. At current rates of compounding progress, sending a complete music CD to friend as an e-mail attachment will, in a few years, require no more relative storage capacity or bandwidth than sending a text file today. A typical home computer hard drive will easily house the entire CD canon in something like 12 years. Nanotechnology has the potential to support this compounding cost reduction for decades.

Thus the moral and legal issues surrounding the digital copying of music may be raised again in broader and broader contexts in the coming decades. Right now it is text, music, pictures, and the like that can be easily copied, but digital descriptions are, in principle, possible for any object. In practice, it will be a very long time, if ever, before digital descriptions of arbitrary objects are possible at the atomic scale. No matter. Less accurate but more compact digital descriptions will be good enough.

An MP3 file "ripped" from a compact disk is a copy of a copy of an original sonic event. The CD format does not digitally encode all the information present at the original event, and for sake of compactness, the MP3 format discards additional information. But once the information is in digital form, regardless of format or fidelity to some original, it can be readily copied an arbitrary number of times with no further loss of fidelity. The fact that MP3 files are imperfect copies does not diminish the intensity of the societal debate regarding their proper use. In the coming decades, an increasing fraction of all the property in the world will be in the form of digital files or will be objects which, at ever diminishing cost or increasing resolution, can be "captured" as digital files. The instruments and manufacturing processes of nanotechnology imply the ability to reconstitute these digital files as "analog" objects when needed. Much good can result from this, and much mischief. It is very important that society get the rules right for the early test case of music.

Privacy and security in a networked world is another current topic of intense public debate. This is another issue that will not be fully resolved for decades as nanotechnology allows the incorporation of information storage, processing, andcommunication in everyday objects at ever decreasing cost. Eventually, nearly every object around us might be networked. The current proliferation of web-enabled cell phones and hand-held computers is just the beginning. We glimpse the future as researchers strive to reduce the unit cost of radio frequency identification (RFID) tags to pennies. Eventually every manufactured object might keep a digital record of its manufacture, distribution, and use.

The benefits have been widely discussed, but the potential for invasion of privacy is clear. Today we fear misuse of data aggregated from our transactions with various web servers. Tomorrow we may fear the misuse of data aggregated from our transactions with our clothing and household appliances. Finally, as a society, we will continue to debate the best ways to ensure broad access to the benefits of information technology.

In our pluralistic society, all viewpoints will be openly and passionately promoted. Current debates over fair use of information technology will generate the initial practices, policies, and laws that will help shape our nanofuture. Unforeseen applications of the technology and new societal issues will continue to arise. But there will be time, and in a democratic society there will certainly be opportunity, to make the necessary mid-course corrections.

Magnetic Storage Applications

Within ten years from the fundamental discovery, the giant magnetoresistance (GMR) effect in nanostructured magnetic multilayers has demonstrated its utility in magnetic sensors for magnetic disk read heads, the key component in a \$34 billion/year hard disk market in 1998. The new read head has extended magnetic disk information storage from 1 to ~20 Gbits/in^2. Because of this technology, most hard disk production is done by U.S.-based companies.

In 3 to 5 years, nonvolatile magnetic random access memory (MRAM) using the giant magnetoresistance phenomenon will be competed in the \$100 billion RAM market. Inplane GMR promised 1Mbit memory chips in

1999. Not only has the size per bit been dramatically reduced, but the memory access time has dropped from milliseconds to 10 nanoseconds. The in-plane approach will likely provide 10-100 Mbit chips by 2002. Since the GMR effect resists radiation damage, these memories will be important to space and defense applications.

The in-plane GMR device performance suffers as the device lateral dimensions get smaller than 1 micron. Government and industry are funding work on a vertical GMR device that gives larger signals as the device dimensions shrink. At 10 nanometer lateral size, these devices could provide signals in excess of 1 volt and memory densities of 10 Gbit on a chip, comparable to that stored on magnetic disks. If successful, this chip would eliminate the need for magneto-mechanical disk storage with its slow access time in msec, large size, weight and power requirements.

Improved Electronic Devices

The integrated circuit (IC) is a main "engine" of today's high-tech/high-productivity economy. During the past four decades, it has continued to revolutionize the ways in which we work, communicate, learn, and are entertained. And this revolution is not over! To the extent that we can continue to miniaturize the components of integrated circuits, we will continue to provide greater functionality and performance at lower power and lower cost. Most of the key IC improvement trends have been driven primarily by scaling to smaller features.

Present integrated-circuit technology spans both the "micro" and "nano" regimes. During IC fabrication, the constituent films are grown/deposited as thin as 2 nanometers and patterned (via optical lithography and plasma etching) into horizontal features as narrow as 100 nanometers (0.1 micrometers). Within the next decade, the pattern dimensions may approach 20 nanometers. Scaling to this level should drive IC manufacturing cost per transistor well below a micro-cent even though cost will rise at the silicon-area level due to the increasing complexity of processing ever-smaller devices and interconnects.

The new electronic products should also uphold their historic trend of "enabling/empowering the individual" — creating new opportunities for high-quality employment and for people to communicate with anyone, anywhere and to access and use vast amounts of information on-demand. Nanotechnology can potentially help extend this vision even further into the future, via both new devices to serve as the "switches and interconnects" of the next IC era as well as by providing a new manufacturing ("nanofabrication") paradigm. Note that some form of (perhaps "bio-based") nanofabrication might be very significant for cost reduction even if applied only to CMOS devices.

If we are successful in developing nanoelectronic devices and nanofabrication techniques for manufacturing them in huge volumes at very low cost, the impact on society will be enormous. New businesses would emerge, many based on the capability to connect almost anything as a node in the "global network". These businesses would help us identify, track, find, safeguard, inventory, control, diagnose, repair, upgrade, etc. virtually any type of item. The full scope of this vision represents arevolution in communication, with a large fraction of all man-made objects brought "online" via at least low-power wireless networking through nearby nodes. There would also, of course, be revolutions in computing based on nanoelectronics. At the high-end, enormously greater parallel processing with much more power at each node would greatly extend our ability to simulate complex systems, such as the weather, protein behavior, etc.

On the portable side, computing at a level beyond today's "high-end desktop" would become available in form-factors limited only by human-interface convenience. For example, the visual display might be "heads-up in eyeglasses" and the input might be primarily voice. The non-I/O hardware would be negligible in size and weight. The entire system, including batteries and short-range wireless data/voice communications, could be integrated, for example, into "small-frame" eyeglasses.

The decreased power requirements enabled by nanoelectronics would be even more significant for this vision than the increased computing performance. We suggest that the supporters of future research in nanotechnology should encourage broad cooperation between universities, industry, and government; emphasize precompetitive results; and include studies on technology choices/down-selection and technology migration/ displacement.

RESEARCH AND DEVELOPMENT

The following potential general outcomes of research in nanotechnology should be of great benefit to society and especially to the electronics and information technology industry:

1. Continued rapid growth of the "high-tech" economy
2. Increased support of university research in the physical sciences, math, and engineering
3. Increased production of work-force educated in the above disciplines
4. Increased opportunity for collaboration between industry, university, and government researchers

5. Significant coordination of government and industry funding and review of academic research, especially through pre-competitive consortia
6. Increased synergy between historically separate research fields

There are more specific potential results of research in nanotechnology that would have a profound impact on electronics and society. We have, perhaps, roughly another decade of rapid advance in integrated circuit technology based on the present paradigm of scaling CMOS. This is indicated in the 1999 International Technology Roadmap for Semiconductors (ITRS), which represents a consensus view on the future technology needs of the industry.

Of course, improved system architectures and design techniques will extend improvement trends at the system level somewhat after the device and process technology mature. Beyond that point, corresponding to lithographic feature sizes approaching the 10-nm scale, fundamentally new types of "switches and interconnects" will probably be required to continue the basic "growth engine" underlying our "hightech" economy.

One of the great opportunities for nanotechnology is to supply such revolutionary electronic components, which would be needed by roughly 2015 at the historical rate of scaling. Carbon (and other) nanotubes quantum dots, and several potential "molecular switches" are all examples of current research that look very promising but need additional development to demonstrate feasibility for nanoelectronics. In addition to the development of new components, nanotechnology holds the promise of a fundamentally new manufacturing process, perhaps based on "biochemical techniques". For nanoelectronics, this could represent the key enabler for continuing the historical decrease in cost per function available through integrated circuits (ICs).

This in itself, even without new types of electronic components, could continue to fuel the high-tech productivity revolution for decades to come.At this point, let's make a quick estimate of the potential impact of nanotechnology on the semiconductor industry via both of the aforementioned mechanisms. The 1999 annual sales of U.S. semiconductor makers was $77 billion, which increased by 4.1x (15% CAGR) as overall CMOS feature sizes were scaled down by 3.9x during the previous decade. Note that overall CMOS feature size is usually characterized in terms of "minimum half-pitch," which is also called the "technology node".

Nanotechnology in the form of a new low-cost manufacturing paradigm could also help to maintain the rapid growth of IC sales even if no new nanoelectronic devices were available to continue IC feature scaling. In fact, reducing cost-per-function is undoubtedly the most important single factor in the approximate proportionality between IC sales growth and feature scaling. Thus, we would expect to see most of the 15% growth rate extended beyond the $300-350 billion level even if we were still making CMOS at a fixed feature size, but with some form of "nano-manufacturing" which continued to provide significant reduction in cost-perfunction.

The main risks for negative societal implications of nanotechnology will probably continue to be in the area of biotechnology rather than electronics. Traditionally, the largest societal risk associated with electronics has been in the area of system/component reliability. Of course, one of the great benefits of microelectronics has been its greatly improved reliability compared to older electronics. And it is expected that further miniaturization of electronics via nanotechnology should result in even further advances in reliability as nanoelectronic systems are designed and fabricated with atomic-level precision.

Of course, the positive impacts of nanoelectronics can also be expressed in more visionary terms. Continuing to improve the cost and performance of integrated circuits through nanotechnology would lead to a future in which almost everything that we manufacture and use could afford to include some electronic functionality. Even for the simplest items, this might include the ability to self-locate/inventory via low-powerwireless communication through neighboring objects/nodes in the ubiquitous "global network."

Wired Society

If we use a length scale defined by a nanometer, nanoscience and technology have been around for several decades, particularly in research, development, and manufacturing in information technology, where film layers and lithographically defined features in the nanometer range are needed.

0It is the wide availability of tools and information, initially produced primarily by the research community associated with information science and technology, to diverse scientific communities, outside of the information related communities, that has generated the current interest in this area. As a result, there is now a very significant broadening of the research base that is interested in nanoscience and technology. A notable example is the involvement of the biosciences community. It is this broadening that makes nanoscience and technology of interest now days.

Diffusion of knowledge, as epitomized by nanoscience and technology, from one discipline to another, so characteristic of all human endeavors, is a reminder that the interface between knowledge and ignorance migrates nonuniformly; rushing ahead in onearea and then spreading laterally to others to produce an apparent broad uniform front, when measured over longer time scales. It is a truism that we cannot predict the future. However, extrapolation of trends in science and technology and lessons from history can be used to guide us about the future.

Computers have a very well defined historical trend. They have been reduced in physical size for a given capability, and their capabilities have continuously improved; these changes have been exponential with time. For example, a computer using vacuum tubes occupied space that was measured in tens of thousands of square feet. This was reduced to thousands of square feet with the introduction of the transistor, and to hundreds of square feet when the first integrated circuits were introduced.

With miniaturization, the smallest commercially available computers have a footprint of some square feet to square inches. In addition, there is now a hierarchy of sizes and distributions, depending upon requirements. It seems to me, by simple extrapolation, that there is the possibility of further shrinking, to a point where computers can be carried by humans twenty-four hours a day.

Similar to, and parallel with, the evolution of computer technology, there has been an exponential change in communication systems. We have gone from fixed sites for initiating communication to the ubiquitous cell phone and personal digital assistant, which provide audio and video information. The networks, which collect information and carry it over distances before aggregation, have also become increasingly hierarchical.

Humans communicate and receive information primarily using audio and visual means. Although tactile senses also provide information. Visual information requires large amounts of data to be transmitted and processed. In the long run this and the need for low-power devices will require the first base station to be not at a distance defined by a kilometer but a meter. In short, relay stations will be ubiquitous in the future. There is some evidence that this is already beginning —the bluetooth technology, for example.Let us examine how these trends translate into some specific possibilities. The human body is already wired by nature. Our nervous system operates on electrical impulses, generated by synaptic connections.

There are many instances in which the electrical signals between the brain and an organ or a part of the body are disturbed or blocked. This could be due to an accident, disease, or defect; nanoscience and technology will play an increasingly important role here. Let us suppose there is a spinal injury, and signals to and from the brain can no longer be transmitted below the injured site. A chip can intercept the neuronal electrical signals, transmit them across the injured site, and then another chip couples the signal back to the body's nervous system. There can be many variants of this idea.

There is already research in this area, but more is encouraged, for the benefits to mankind are obvious. The next level of complexity involves nanoscale devices that can sense human temperature, pressure, or blood chemistry. These devices transmit, wirelessly, on a continuous basis the state of the human body. As long as the readings are within an acceptable level for a particular individual, no medical intervention is called for, but if they deviate from a defined range, intervention, by a human or a machine, becomes an option. There are already many ideas of using chips that can deliver drugs to local sites, and that bond, say, to a tumor and signal its whereabouts, and are wirelessly instructed to deliver a drug. Clearly, success of these possibilities can have profound implications for health care.

Humans have an innate tendency to prefer not to carry devices or be bound by them. For example, the wireless telephone is preferred over one that needs a chord. It follows then, if one did not have to carry a personal digital assistant in one's hand and peer at a small display that would be even more desirable. If communication relay stations are all over, say within a meter range, then very low-power transmitting and receiving devices can be built. These can be carried or even implanted into adult humans. We know microphones can be implanted into throats, as is in the case of patients with cancer. Similarly, ear implants can be directly coupled to the mechanical sensors in the inner ear.

More problematical is visual information. Carrying large displays is a nuisance. Small projection displays have their drawbacks, similar to those of headphones. Perhaps a nanoscanning device, which directly projects a rastered image onto the fovea centralis, rather than the full retina may be invented. A ray of light will require compensation for refraction, if it has to traverse the cornea, but may not if it proceeds only through the vitreous medium. There is considerable room here for invention. If all of these speculations come to pass then a human will be wired fully — not only internally but also externally to the vast network outside of the body.

Quantum Legos

Our desire to understand nature has always been guided by a philosophy of reductionism. In order to comprehend the functioning of the very diverse world of nature, we invented the disciplines of physics, chemistry, materials, biology, etc. These partitions of nature have served us well. We are able to train students, develop methodologies, and perhapsmost importantly, research subjects that interested us as individuals.

The disciplines provided us an umbrella at academic institutions, within which we could advance our field, belong to, and be appreciated by. There was a similar evolution of disciplines in engineering but with a very different outlook. The engineers were not so much concerned with understanding the laws of nature but rather in using them to build something useful for mankind. In contrast to the reductionism of scientists, engineers are synthesizers.

Underlying most "hard" sciences are a handful of atoms. They combine in a myriad of ways to produce the world around us and including us. The physical scientists have worked hard to explain how things around us can be understood in terms of collection of atoms and molecules. Within the last decade or so, following the discovery of the scanning microscope and atom trapping devices, we have begun to "play" with single atoms or molecules; these have length scales at the low end of the nanometer and are quantized in their behavior. It is only a matter of time when these "quantum legos" will evolve into a (new) field of quantum science and engineering. Researchers working in this field-to-be are not likely to view their working philosophy as reductionism. They will synthesize new arrangement of atoms, learn ways to replicate them, and produce objects of interest to mankind.

One day we will synthesize or build a molecule that, like the DNA, will store a code that leads to the production of material objects. As in the case of nanoscience and technology, quantum science and engineering has also been around for a number of decades. It has not received wider recognition because it was also primarily confined to the information technology industry. It is used, for example, in the operation of transistors or lasers.

.Quantum mechanical considerations can be essential in many systems, where the number of atoms or the length scales exceed the nanometer range, for example, in superconductors or in photonic systems. Quantum cryptography, teleportation, or computers also fall into the broad field encompassed by quantum science and engineering.

There will be numerous important contributions made by nanoscience and technology in many diverse areas of technology that benefit mankind. Humans have always had the desire to generate, diffuse, and receive information about each other and the universe around them. What is changing very rapidly is the rate at which these three components are being implemented, the ease with which this can be done, and the nature of information. This virtual ocean of information about every aspect of life, including health, financial, and personal behavior, increasingly surrounds us.

Science and Education Implications

Nanotechnology is a current "New New Thing" in science: an area that promises new understanding of nature, and use of that understanding to build technologies that will change the world. It has captured the attention of the public and of the government, and is beginning to attract the attention of corporations. Because it is new and exciting, it has also caused flurries of public concern.

There is no doubt that nanoscience is a breath takingly interesting area, and one of the most exciting in modern physical science. It is also clear that the path from science to technology is just beginning to emerge, and it is too early to say whether the impact of nanotechnology on society will be revolutionary or insignificant.

Nanoscience is the study of systems with nanometer dimensions. The upper limit of size of a nanostructure is often taken as 100 nm, but truly new phenomena —

quantum behavior, properties more closely resembling molecules than microscopic objects — usually occurs at much smaller scales: 1-10 nm. Highly developed nanostructures are most evident in biology, where a host of important subsystems — from ribosomes to viruses — have nm-scale dimensions. Chemists, working from the "bottom up" have made structures — molecules — at the bottom end of this scale for many years, and have developed great skill in placing atoms in structured aggregates with great precision. They have only recently begun to connect this synthetic technology to interests in microelectronics and related areas.

Solid-state science has historically worked from the "top down", by writing functional patterns into macroscopic sheets of semiconductor. One of the hopes for nanoscience and technology is that the combination of a number of areas — from both "top-down" and "bottom-up" fields, and from biology and computer science — will create a new area and lead to major advances in both understanding of science and in applications of science in technology.

Scientific Opportunities

Quantum Phenomena. Nanotechnology will be an important part of any effort to exploit quantum phenomena. Quantum behavior — especially at room temperature — depends on small structures. Especially for microelectronics — where the wavelength and ballistic mean free path of electrons are crucial parameters — room temperature quantum behavior becomes important only at dimensions of a few nanometers. Developingscience and technology that explores and exploits these behaviors will require developing methods for fabrication that operate on that scale.

There are now a number of demonstrations of quantum behavior in small systems, with examples ranging from "quantum corrals" to fluorescent quantum dots (and, of course, with many of the properties of molecules determined by quantum behavior); there is every reason to expect many more to emerge, since quantum behavior dominates small structures.

Science of Large Numbers of Objects. Phenomena that involve very large numbers of small components follow different rules than those involving only a few components: collections of apples do not behave in the same way as collections of molecules. Nanofabrication offers the potential to synthesize small objects in very large numbers, and thus to build systems in which it is possible to examine forms of behavior that are difficult to observe when only small numbers of components interact: very large combinatorial experiments, experiments in directed evolution, and information/ computation systems based on cellular automata are examples.

Atomic- and Molecular-Scale Structure and Fabrication. The combination of molecular synthesis and advanced lithography offers a unique opportunity to fabricate nanostructures with high degrees of control. Chemistry — that is, molecular synthesis — already has enormously sophisticated tools for building highly structured collections of atoms, and chemistry routinely fabricates sophisticated nanostructures — that is, molecules — one atom at a time in quantities of tons. What chemistry has not provided is methods of building these structures in ways that are electronically or optically functional in ways that are analogous to those required in computation or telecommunications. Advanced lithography can manipulate materials relevant to electronics, but is cumbersome when applied to structures with dimensions below 100 nm. The opportunity to combine the sophisticated techniques of these two areas is one of the most exciting opportunities in nanofabrication.

Nanoscale Materials: Quantum Dots, Magnetic Materials, Buckytubes, Others. One of the most rapidly emerging applications of nanotechnology is the fabrication of functional structures with nanometer dimensions. Carbon-based systems, semiconductor colloids as fluorophores for bioassays, shaped semiconductor nanocrystals, FePt and Co nanoscale particles for magnetic information storage — all are current examples. There will almost certainly be many more to come. The techniques of small particles — nanocrystals, emulsions, colloids, micelles — are highly developed in chemistry and materials research, but have not been applied to materials that can be integrated into electronic and optical systems.

The Cell. Understanding the cell is one of the great challenges of biology, and of modern science. To understand the cell, it will be necessary to examine it using probes that are small enough that they do not perturb it. Since mammalian cells are usually tens of μm in size, and bacterial cells are usually 1-5 μm in size, probes should be small on this scale. Nanostructures will almost certainly have important applications in studying cells — as intracellular probes, as tools for stimulating the cell and for measuring its responses electrically, magnetically, and optically, and in devices for manipulating the cell in sophisticated ways.

Tools and Understanding

One of the reasons that nanoscience is a rapidly expanding area is the rapid emergence of tools for study of nanoscale phenomena, and for fabrication of nanoscale structures.

Scanning Probe Microscopies. The scanning probe microscopies — of which there are now many variants —

have revolutionized nanoscience, by providing the highestresolution methods now available for determining the structures of surfaces.

Single-Molecule Studies. Methods for examining the behavior of single molecules are exploding. The most highly advanced are fluorescence-based optical methods, and scanning microscopic methods, but others — especially SEM — are also useful.

Nano-Scale Materials. Synthetic methods leading to nanostructures have been available for many years in chemistry and materials science, but the fact that these methods provided nanostructures was, in a sense, often incidental to other objectives. It is now appropriate to consider these methods as a treasure of highly developed synthetic methods capable of being applied to current objectives in nanoscience.

Atomic-Level Structure: Electron Microscopy, x-ray Diffraction, Nuclear Magnetic Resonance Spectroscopy. A variety of techniques are capable of supplying atomic-level information about structure in appropriate systems.

Molecular Synthesis. Molecular synthesis has become an enormously sophisticated field, with the ability to build an enormous diversity of molecular structures. The technology of chemical synthesis is available for use in the synthesis of nanostructures.

Meso-scale Synthesis. Extension of concepts from molecular-scale synthesis to the assembly of aggregates at larger scales is just beginning, but shows great promise. The assembly of crystalline lattices of ~100 nm-scale spherical beads is routine, although the number of structures is limited. Two-dimensional structures with high order — liquid crystals, self-assembled monolayers, colloidal crystals — are examples of ordered structures.

Biology: Genomics, Proteomics, and Cell Biology. Molecular biology has been the field that has probably experienced the most revolutionary change in the last 50 years. Biology now provides an almost unlimited range of nm-scale problems, a growingnumber of examples of nm-scale phenomena, and a need for nm-scale probes and devices. The opportunities and problems in biology and biochemistry range from molecular issues to larger scale problems.

Technological Motivation

The opportunities in nanostructures are very rich, when viewed from the vantage of science. The arguments are equally compelling from the point of view of technology. *Extensions of Moore's Law.* The continuing decrease in the size and cost of transistors, and the corresponding increase in the their density, in the capability of the devices that are fabricated from them, have been major driving forces for the economy. There is general concern that current technologies are "running out", and that the rate of technological advance will be slower in the future than in the past.

Microlithography using light and electrons has shown a remarkable ability to make smaller and smaller structures, and there is no reason to consider that the decrease in size of the last decades will not continue, at least for a while. Smaller sizes will, however, be more and more expensive. Moreover, as devices approach sizes of 20-50 nm, new materials issues become important. Understanding what is and is not possible in nanoscale information processing and storage devices is a key technological imperative. If it proves to be possible to go to substantially smaller devices with acceptable quality and cost, the devices fabricated using those technologies will have global impact.

Quantum Technology. A number of applications for nanoscale structures exist. Semiconductor quantum dots are being used very successfully as fluorescent labels for biological systems, and have the advantage — relative to molecular fluorophores — that they resist photobleaching. Superparamagnetic particles are used for contrast enhancement in clinical applications of magnetic resonance imaging; different compositions have been proposed for use in information storage. With small size unquestionably comes new materials properties, and the potential for new applications. The search for real applications of nanotechnology continues, however, because the field has not yet identified areas in which it can contribute unique technologies.

Biomimetic or Biologically-Based Devices. Biology provides an enormous stimulus for nanoscience. The cell is based on the operation of "nanomachines" of many types — from relatively simple catalysts (enzymes) to much more complex systems — the ribosome, the light-harvesting apparatus of photosynthetic plants, the Golgi apparatus, components involved in DNA replication and in mitosis, the rotary and linear motors that are ubiquitous in cells.

The significance of these biological systems for the development of nanotechnology in the short term is not entirely clear. At worst, they provide existence proofs for molecular machines so much more sophisticated than we can presently design that simply understanding theirprinciples of operation will keep biochemists and nanoscientists occupied for decades.At best, they may provide suggestions of designs for nanodevices, strategies for using nanostructures in new types of functions, and perhaps even components for new types of devices.

Databases and Information Storage. One of the early types of devices to reach the prototype stage in nanotechnology are very high-density information storage devices. The most advanced of these systems are CD-like devices in which the size of a bit is ~50 nm, and in which writing, reading, and erasing are achieved with the tip of a scanning probe device. The use of nanostructured magnetic materials also offers the possibility of very high densities of information. These sorts of systems will make it routine to store very large quantities of data, and to be able to search them efficiently.

Portable, High-Performance Systems. With very small microelectronic systems would come the ability to fabricate new classes of devices, and to revolutionize a number of functions. The movement from large computers to desk-tops, and from desk-tops to laptops and personal assistants, have fundamentally changed the way that data is used. If the power of the PC or the local server could be put on the wrist or worn on a belt, the world would change again. Nanotechnology — with its potential for making functional systems that are very small — has the potential to make key contributions to the goals of dramatically reducing size, and decreasing power consumption.

Sub-Wavelength Optics. A simple type of application of nanotechnology is in the area of sub-wavelength optical devices and systems: that is, devices and systems that manipulate light using structures substantially shorter than the wavelength of this light. This type of technology has the opportunity to make a range of devices — from optical filters to switches — in nanometer-scale systems having good performance and low cost.

Functional Materials: Catalysts, Materials for Energy Storage. Nanostructures are already an important class of material in many area of energy production and storage: catalysts are often based on chemically functional nanostructures; zeolites are structures with nm-scale internal dimensions, and catalysts often have nm-scale features. Bringing these classes of structures under better control, and developing the ability to design catalytic activity, has been an ambition of this field for a decade. Although progress has been spectacular, there are still major hurdles to be leapt before rational design of catalysts and other energy-related materials become a commercial reality.

Seizing The Moment: Education And Training. It is clear that there are both major scientific opportunities in nanoscience, and major requirements for nanostructures and devices from nanotechnology. What will be required to convert science into technology in this area? The first, and key issue, is the availability of scientists and engineers able to work effectively in this new discipline. This group must also have the financial support necessarily to carry out their work efficiently. It is possible but not optimal to rely on the current system of education and training to produce nanoscientists and engineers; the characteristics of the field require, optimally, a different type of training.

Nanotechnology Spans the Conventional Disciplines. Nanoscience and technology do not fit within any of the conventional scientific disciplines. Scientists and engineers developing nanoscience and technology will have to be aware of a broader and different range of subjects than those included in the usual departmental curricula. For example, a scientist interested in nanoelectronic devices should be aware of the types of functional nanostructures found in cells, and familiar with the concepts and capabilities of colloid chemistry and self-assembly. Developing the educational system to educate and train in these areas will require substantial changes in the university, with a shift from disciplinary to multidisciplinary education being an important strategy.

Collaborative Research. Since the problems in nanoscience and technology will require a broad range of talents, collaborative research will probably be more the norm than the exception in universities, and perhaps in industry. Developing styles of research that span multiple disciplines is a growing skill in universities, but one whose importance needs constant reinforcement.

Shared Facilities. Nanoscience is "intermediate" science. It does require sophisticated physical measurement, and access to specialized devices for fabrication, but generally does not require very large facilities of the types represented by synchrotrons. Certain types of equipment — particularly those used for nanofabrication using e-beam writing and related techniques, and highly disciplined cleanrooms — probably do require regional facilities. Much of nanoscience is, however, done in university laboratories, and a scanning probe microscope that is local probably has more impact on the rate of progress of the field than an electron microscope located in a different state.

Types of Training. Because nanoscience and nanotechnology are interdisciplinary, the time require to learn the basics of the field may be longer than that for single disciplines, and postdoctoral training may be more a more important part of the field than it is in conventional physics, chemistry, or materials science. At the other end of the scale, countries such as Switzerland have recognized the importance of well-trained people at a lower level — master technicians — in moving a new technology into commercial reality. To anticipate the need for skilled technicians, the Swiss system of training includes programs explicitly focused on this group.

Future Outcomes

There is no question that nanoscience — the exploration of phenomena at the nanometer scale, and the generation of understanding of these phenomena — is exploding. The potential formovement of that knowledge into practical application is the basis for the popular enthusiasm for this area, and the anticipation presently has reached beyond the ability of the field to deliver. For the health of the area, it is important that the scientists not promise too rapid emergence of technologies from their efforts. There are many delays between science and technology, especially in areas where manufacturing requires the production of products of very low cost and very high reliability.

Nanomaterials — functional colloids, materials such as buckytubes, quantum dots, phase-separated copolymers — are already beginning to be an important part of materials science, and the first applications — of gold and cadmium selenide nanodots as parts of clinical and biochemical research reagents for example — are already established. The future will undoubtedly see the rapid growth of the area of nanosynthesis of small (<20 nm) structures using the techniques of synthetic chemistry, and the development of applications for the resulting structures. These systems will make room-temperature quantum behavior widely available, albeit in systems that are only isolated particles or collections of particles that interact only locally.

Self-assembly will play an essential role in converting nano-scale particulate materials into useful components of more complex systems: positioning the very large numbers of individual particles that would make up such systems using robotics or some other external agency is only slightly less impractical than positioning molecules in a molecular material.

An early application for new new science is often in the development of tools that enable further investigation and measurement. These tools are already beginning to emerge rapidly from nanoscience. The family of scanning probe devices is one obvious example: these devices are now making the transition from pure research tools to tools for industrial metrology, and in that capacity will allow the measurement of characteristics of manufactured systems with a precision that was unimaginable 20 years ago. Fluorescent quantum dots are being explored actively as color labels for cell biology, and will probably move rapidly into clinical diagnostics. Buckytubes are the subject of large-scale investigation, and one of their first applications is in the fabrication of new, high-resolution probes for scanning probe devices.

A significant part of the enthusiasm for nanotechnology is based on the possibility that it will extend Moore's law for microelectronic devices to dimensions that cannot be reached economically by existing technologies for microfabrication. The apparently inexorable progression of transistors toward smaller features has already produced conventional technology that is close to production-ready at <100 nm design rules, prototypes of transistors using extensions of this technology that have gates of 20– 30 nm, and devices that can operate on the basis of the movement of a single electron. Nanotechnology will, it is widely hoped, play an important role in moving these devices toward commercial reality. The development of a large-scale commercial technology from one-of-a-kind prototypes is an expensive and long-term process, but the history of technology suggests that such laboratory demonstrations anticipate commercial reality surprisingly often: what can be made one, can often be made better in multitudes.

The issues in building nanoelectronic systems are, of course, much more complex than just building small transistors. Wiring these transistors into functioning systems is onemajor challenge; a second is managing the interaction between devices; a third is designing systems that will adapt to the almost inevitable failure of individual devices that will occur in systems containing numbers of devices that are presently unimaginably large.

The immediate outcome of success in making information processing and storage systems on the nanoscale would be to make levels of computing power that are presently unimaginable available for tasks that can now only be imagined: machine translation of complex spoken or written text, weather and economic prediction, data mining of enormous stores of data.

While chemical synthesis has been developed to a breathtaking level of sophistication, and while the ability of chemists to position individual atoms in exactly defined positions in very complex structures is established technology, the combination of this type of expertise with areas of technology commonly considered as nanotechnology — especially that for computation and information storage — has just begun.

The area of single-molecule electronics may be the precursor of such a combination. This area is exploring the proposition that complex molecular structures might become the basis for information systems: a molecule might, for example, become a transistor, switch, or memory cell. There are both practical and conceptual problems with this vision of a future nanotechnology, and its long-term impact on real computing systems cannot presently be evaluated. Nevertheless, it seems certain that as dimensions shrink toward the molecular, and as conventional "top down" fabrication becomes exponentially more difficult, synthetic, "bottom up" approaches will grow in importance.

Policy Issues

Nanoscience and nanotechnology is a new area that offers access to structures, properties, and behaviors that have not previously been accessible. As with all new areas of science, with the opportunity comes challenges in policy.

Public and Private Financing. The revolution in microelectronics was financed with a combination of public and private investment. Much of the early development of molecular biology — the precursor of biotechnology — was carried out in universities with public support. Much of the early development of polymer science was carried out in industry, with private financing.

International Competition. Nanotechnology has become a target of greater or lesser interest to virtually every industrialized nation. Among those regions that are active in local programs in nanotechnology are western Europe (where a number of countries have active private and public programs), Japan (where it is an obvious extension of the Japanese strength in microelectronics) and Israel (which has a national enthusiasm for high technology that has deep roots in concerns for economic security). It is useful to remember that the most important inventions of nanotechnology scanning probe microscopy, C60, giant magnetoresistive materials.

Implications for National Security. A number of areas relevant to national security have already been identified as targets for nanotechnology. Quantum computing is one possible approach to important problems in cryptography. Global information systems for the DOD will require the movement and analysis of staggering amounts of information, and the development of new tools for information management and data mining.

Privacy. The potential of nanotechnology to make it possible to acquire and manipulate very large sets of data raises fundamental questions about privacy. When it is possible to track every citizen, and to store information that would predict patterns of future behavior, it is important to set guidelines for the use of that information early. The development of large commercial databases using current technology is proceeding rapidly; nanotechnology has the capability to accelerate this development in the future.

Public Perception: "Grey Goo". As with many new technologies, nanotechnology is the subject of public concern. Encouraging and building a sound public understanding ofthe potentials and limitations of nanotechnology is essential in avoiding future misunderstandings based on misperception.

Bibliography

Delzeit, L., C. V. Nguyen, B. Chen, R. Stevens, A. Cassell, J. Han, and M. Meyyappan, "Multiwalled carbon nanotubes by chemical vapor deposition using multilayered metal catalysts," *J. Phys. Chem. B*, vol. 106, 2002.

Feynman, 1. R.P. "There's Plenty of Room at the Bottom," *Engineering and Science.* Caltech, 1960.

Hornbogen, E. "Precipitation Hardening—The Oldest Nanotechnology," *Lightweight Alloys for Aerospace Application*, ed. K. Jata et al. Warrendale, PA: TMS, 2001.

Huang, Z. P., J. W. Xu, Z. F. Ren, J. H. Wang, M. P. Siegal, and P. N. Provencio, "Growth of highly oriented carbon nanotubes by plasma-enhanced hot filament chemical vapor deposition," *Appl. Phys. Lett.*, vol. Dec. 1998.

Iijima, S. "Helical microtubules of graphitic carbon," *Nature*, vol. 354, 1991.

Jones, R. "The Future of Nanotechnology" August 2004.

Kittelson, D. B., "Engines and Nanoparticles: A Review". *Journal of Aerosol Science* 1998.

Murr L.E. et al., "Carbon Nanotubes and Other Fullerene-Related Nanocrystals in the Environment: A TEM Study," *JOM*, 56 (6) 2004.

Peters A. Wichmann H. E.; Tuch T.; Heinrich J. and Heyder J. "Respiratory Effects are Associated with the Number of Ultrafine Particles", *American Journal of Respiratory and Critical Care Medicine*, Vol.155, Issue 4, 1376-1383. 1997.

Reynolds J.G. and B.R Hart, "Nanomaterials and Their Application to Defense and Homeland Security," *JOM*, 56 (1) 2004.

Rittner, M.N. "Nanomaterials in Nanoelectronics: Who's Who and What's Next," *JOM*, 56 (6) 2004.

Roncone, K. "Nanotechnology: What Next- Generation Warriors Will Wear," *JOM*, 56 (1) 2004.

Sioutas, Li, N., C., Froines, J.R., Cho, A., Misra, C and Nel, A. "Ultrafine Particulate Pollutants Induce Oxidative Stress and Mitochondrial Damage" *Environmental Health Perspectives*, 2003.

Stevens, J. Li, R., L. Delzeit, H. T. Ng, A. Cassell, J. Han, and M. Meyyappan, "Electronic properties of multiwalled carbon nanotubes in an embedded vertical array," *Appl. Phys. Lett.*, vol. 81, pp. 910–912, July 2002.

Index